ANIMALS OF THE AMERICAS

Animals of the Americas

DR JIŘÍ FELIX

ILLUSTRATED BY KVĚTOSLAV HÍSEK, JAROMÍR KNOTEK,
LIBUŠE KNOTKOVÁ AND ALENA ČEPICKÁ

HAMLYN

LONDON • NEW YORK • SYDNEY • TORONTO

Designed and produced by Artia for
The Hamlyn Publishing Group Limited
London · New York · Sydney · Toronto
Astronaut House, Feltham, Middlesex, England

Translated by Dana Hábová
Graphic design by Vladimír Šmerda
Photographs: Miloš Hornof (p. 102), Oldřich Mazurek (p. 188),
Ivo Petřík (p. 228, 270), Vladimír Plešinger (p. 12, 146)

ISBN 0 600 36643 X

Printed in Czechoslovakia by Svoboda, Prague
1/12/05/51—01

CONTENTS

INTRODUCTION

The time is long past when pioneer wagons rumbled across the North American prairies. Today, there is a network of highways. In Alaska, dog sleds are no longer a means of transportation in a land where tracked vehicles abound. Today, various means of transport are available to take us to the most inaccessible area of our planet, the South American rainforests. Few people nowadays think of this continent as the New World. And yet, it remains a land of many mysteries, not only because it still hides places that humans have never entered, but also because even in ancient settlements, nature has kept many of its secrets, and has many a surprise in store for us. By way of the pages of this book, you are invited to take a trip into the animal world, which may well be a new world for you.

You will visit a continent of extremes. It is the home of the tallest trees and the smallest birds, of the largest carnivorous mammals and the tiniest predatory fishes. It is the continent with the most extensive protected natural reserves and parks, and it is also the continent where man has invaded the natural environment more cruelly and mercilessly than anywhere else in the world.

You will meet many members of the huge family of animals which inhabit the American continent. Let us recall the fate of horses in America. They were first introduced by conquerors from Europe. Some escaped from captivity and multiplied in the wild, so that herds of mustangs became widespread on the American plains. Later they were killed, until there were only a few left. Only then, at great expense and with much effort, did the nation try to save this species for future generations. In this book, there are many similar examples of man's contradictory behaviour towards wildlife, and they are part of the study of American fauna.

The whole continent, covering some forty-two million square kilometres and ranging almost from Pole to Pole, bears the name America. Its fauna, however, consists of several different, markedly separate worlds. The most obvious difference exists between the northern and southern parts of the continent, and this is not only because of the different climatic zones. The two parts of the now united continent first emerged from the ocean, far apart from each other. Present-day North America was joined to Europe, which accounts for the many characteristics common to the fauna of these continents. The fauna of South America is very markedly different. South America gradually approached the ancient continent and formed one with its northern part. This happened relatively recently in geological time — about 250 000 years ago. Even though we find it difficult to imagine such a length of time, it is but a brief part of the evolution of natural life, and differences established before the continents joined are still very much in evidence.

Let us set out to investigate the fauna of North, Central and South America. It will be a journey of discovery and of mystery and perhaps even of adventure. But we shall not follow the path of the hunters, adventurers, gold-diggers and men with guns. We shall follow the scientists and explorers. We shall see the result of their work in hundreds of pictures and read about them, taking part in a great adventure, called the adventure of knowledge.

We believe it will interest you, teach you something new, and give you pleasure.

THE FORESTS OF SOUTH AND CENTRAL AMERICA

The lush green world of the primary forests of South and Central America is one of the least accessible areas of our planet. Here millions of square kilometres are covered with thick jungle, not easily penetrated even with the aid of modern technology. Long ago, when the first discoverers tried to probe the primary forests and conquer the land, their efforts were about as effective as trying to cut down an old oak tree with a penknife. The rainforests of the Amazonian region alone cover a larger area than those of all the other continents put together.

Gradually however, this fortress, constructed over thousands of years by tropical forces, yielded to the assault of the more determined invaders. They used the many wide and navigable rivers to force an entry, plundering the natural riches of the country — gold and silver, rubber, furs and rare birds' plumage. Over the years many species were lost for ever in order to satisfy the ever-increasing demand for such things as fashionable hat decorations . . . But others, less demanding, came to the primary forests — explorers and scientists seeking different treasures — the bounty of hitherto unknown plants and animals.

The task was never easy. The climate alone is difficult for foreigners, especially Europeans. Although a tropical rainforest maintains a relatively uniform temperature of about 27° C, there is a high degree of humidity and most visitors find the steamy heat almost unbearable.

Moreover a newcomer may well be intimidated, not only by the size of the forest, but also by the many inexplicable and often frightening noises. From dawn to dusk, the cries of myriads of songbirds, colourful parrots, toucans, monkeys and other creatures can be heard. Then, with surprising suddenness, the tropical twilight sets in, and the forest vibrates with the hooting of owls, the howling of beasts of prey and the croaking of frogs. And there are other unfamiliar sounds to terrify the traveller. Explosions like cannon fire may echo through the forest, alarming both newcomers and natives. The South American Indians used to say, 'Ghosts are roaming in the darkness . . . ', but it is only the cracking of the fruits of the Sandbox Tree *(Hura crepitans),* often called the cannon tree. When the fruit is ripe, the tough, woody outside layers separate suddenly from the core, scattering seeds with the explosive noise of a hand grenade, and dispersing them far and wide.

Present-day travellers are usually well informed, and realize that every frightening experience is both natural and capable of explanation. Nevertheless, the unexperienced visitor still fears such creatures as snakes, knowing that some local species are among the most venomous in the world. A stranger may only slowly overcome his many prejudices and superstitions. He has to learn that no wild beast attacks without reason, and no snake will kill if it is left undisturbed. But many snakes are dangerous. They often enter human habitations at night in pursuit of their prey, and if inadvertently touched, a snake will bite in self-defence. Then every second counts, for an anti-snakebite serum must be administered immediately if the victim is not to die. Thousands of people used to perish every year as a result of snake bites, but nowadays, if efficient serums are used in time, many lives can be saved. There are several large snake farms, where researchers seek effective antidotes. One of the best-known farms is Butantan in São Paulo in Brazil. Here specimens of venom are received from snake hunters in every part of South America, and the Butantan serums save lives throughout the continent.

However, it is not snakes or preying animals which are the greatest plague of primary forests. A small spider, called the black widow, still causes genuine alarm for its bite is just as dangerous as that of a snake. Fortunately efficient antidote is now being produced. Mosquitoes are another ever present cause for concern. Milliards of them flourish in the warm humid conditions, and there is no escape from them. Their constant attacks and painful bites can drive a person mad, and their bite may introduce the viruses of malaria and yellow fever into the bloodstream of their victims. Both diseases were once fatal. Modern medicine has developed efficient remedies, but mosquitoes are still among the worst pests of tropical forests.

There is another mysterious forest phantom — the vampire bat, whose name has become associated with horror fiction. It flies silently out of the darkness, and disappears just as noiselessly, leaving behind it only a slight wound, which cannot, at first, be accounted for.

Primary forests hide many treasures and many raw materials, some of them irreplaceable. They gave the South American Indians a most fearsome weapon — curare. It is poison of plant origin, used as a coating for arrows and blowpipe darts. Blowpipes are effective up to about forty metres, and have the advantage of being completely silent. When the first intruders penetrated the territory of the South American Indians, the warriors attacked them with poisoned arrows which swished silently from the green jungle and caused a horrible death, similar to that brought about by tetanus. Curare was for a long time one of the terrifying mysteries of the forest, and the Indians guarded their secret well. After many years of exploration and research, its source was found in the juice of a climbing liana with deltoid leaves and clusters of purple blossoms, called *Chondrodendron toxicoferum.* Together with the juice of the tree *Abuta rufescens* and other vegetable extracts, the liquid was thickened by heating, and changed into one of the most dangerous nerve poisons ever discovered. Nowadays curare is widely used to relax muscles and nerves, both in surgery and in the treatment of many ailments.

The Cocoa Tree *(Theobroma cacao)* is a native of the South American rainforests. It has pink flowers and large red or yellow fruits on both the branches and trunk. The pods are 15 — 20 cm long, and contain cocoa beans. These are allowed to ferment before being dried. They are later roasted, shelled and powdered to give the well-known product. Cocoa trees were originally found wild in the forests of South America, but they are now cultivated in plantations.

Very often, though we may not realize it, we can smell the Mexican orchid *Vanilla fragrans,* whose pods give us the familiar spice called vanilla.

The rainforests of the Amazon basin in Brazil gave the world a plant, the juice of which is used throughout the world — the rubber tree. This is the name given to many species, such as *Hevea guianensis* or *Hevea brasiliensis,* and trees of the genus *Manihot,* whose bark, when cut, produces raw, milky latex.

There is an inestimable number of plants and animals, many still unknown, in the tropical forests. The classified trees alone exceed two and a half thousand. The native population knew how to make lethal poisons out of their fruits, bark and juice, but they also used them as staple foods and delicacies, and as medicines. South America is the home of the pineapple, and of the cinchona tree from the bark of which quinine is obtained. This alkaloid cures fevers and malaria. From the forests also come timber, fragrant oils and nuts.

The South and Central American primary forests have been called a 'green hell', but to many who know them intimately they are a paradise — magnificent, dazzling, colourful and teeming with life. In this book they will be neither a hell, nor a paradise. We shall speak about what is known, remembering, that in no continent are there so many unknown species among the flora and fauna. The number of known insects which live there is in the region of hundreds of thousands, and how many more unknown species there are, it is impossible to tell.

It was difficult to decide which animal species to include in this book. It should be recalled that a tropical rainforest is an intricate community of plants and animals, its balance assured by many mechanisms. The most important is the division of its space into several storeys, as in a modern building, each storey inhabited by a group of animals, and providing food and shelter for them and their young. Each has its predators and parasites, helping to maintain the biological balance and preventing one species from taking all the food. We must not forget that when we speak about an inhabitant of the forest, there are always others in the neighbourhood. And that applies to life elsewhere than in a forest.

And now, let us set out on our journey . . . .

AZARA'S OPOSSUM
Didelphis azarae

Azara's Opossum is an inhabitant of forests and scrubland in tropical South America. It is about 70 cm long, including the tail, which measures about 30 cm. It possesses a large number of teeth, having 50 altogether, of which 10 incisors are situated in the upper jaw and 8 incisors in the lower. No other mammal has so many incisors. Another peculiarity is its prehensile tail, which enables the opossum to hang on to branches, or to pluck twigs and leaves for building its den. The material is carried in the curled up tail. The opossum makes its den in a hollow tree, in a rocky crevice or in a thicket. It may settle in the roofs of storehouses near forests, or even in sheds and barns on the edges of villages. It visits hen-houses and often causes widespread destruction at night, killing more chickens than it can eat. The opossum seldom devours its prey, preferring to lick the fresh blood, so in the morning, the farmer finds his flock reduced to mutilated carcasses. No wonder then that poultry farmers hate this predatory marsupial and kill it on sight. It is also sought for its fine fur, which is made into expensive collars, hats and coats, and the native Indians hunt it for its tasty meat.

The opossum feeds on birds, small mammals, lizards and amphibians, as well as on insects and other invertebrates. It hunts after nightfall, staying in its den by day. It does not have a permanent home and often changes its den. Before sleeping, it rolls and folds down its large ears to shut out the noise of the forest, for it has a very keen sense of hearing, and would be disturbed by the sounds around its shelter.

The opossum is solitary, seeking a mate only in the breeding season. After the very short gestation period of 12 — 13 days, the female gives birth to 4 — 11 young. They are at an early stage of development, are naked and blind, and are only about 1 cm long. They immediately crawl through their mother's fur to her pouch and attach themselves to her nipples. They suck the milk for about 70 days, then leave the pouch and ride on their mother's back for several weeks, holding on to her tail with their own tiny tails. After weaning, the young eat prey brought by the female, and become independent at the age of 15 weeks. Females mature after 24 — 26 weeks, and males after 32 — 35 weeks. Young females produce young once or twice a year, while older females have three or four litters annually. Although they produce young in great numbers, these marsupials have many enemies, falling prey to birds, owls, carnivores and snakes. Their sharp teeth are little help, so when threatened the opossum feigns death, lying stiff with its tongue hanging out. As most carnivores prey only on live game and ignore carcasses, this instinctive behaviour often saves the opossum's life.

FOUR-EYED OPOSSUM
Metachirops opossum

The Four-eyed Opossum does not have four eyes, but obtained its nickname because it has two round white patches above its eyes. It is 50 cm long, and is found in tropical regions of Central and South America. It is resident in the forests where it moves with agility among the trees. It catches insects and small lizards, especially geckos, in the treetops and on the

Azara's Opossum

Four-eyed Opossum

trunks, and it takes birds' eggs and nestlings. It also feeds on sweet fallen fruit. It forages after nightfall, hiding by day on thick branches, in dens made of leaves and twigs. The ball-shaped den has a side entrance.
After a gestation period lasting about 12 days, the female produces an average litter of 6 blind and naked young. They continue their development in their mother's pouch, sucking milk from nipples to which they are attached. They leave the pouch after two months, and then ride on their mother's back for some weeks longer.

Mouse Opossum

MOUSE OPOSSUM
Marmosa murina

The Mouse Opossum is the smallest American marsupial. It is 25 cm long, but 15 cm of this is taken up by the tail. It is found in the tropics of South and Central America and on the adjacent islands. Unlike other species of opossums, the female Mouse Opossum has no pouch, and the mother carries her young attached to her abdomen. The young are born naked and blind after a gestation period of 12 days. They remain attached to their mother's nipples for about 8 weeks, before they start crawling over her body, and stay for 2 weeks on her back, coiling their long prehensile tails around hers.
The Mouse Opossum is nocturnal. During the day, it shelters in hollows but it comes out after sunset to seek its food, hunting insects, spiders, and other small invertebrates. It also catches small birds and takes their eggs.
It lives to be 5 — 6 years old. It is sometimes found in North America and even in Europe. This is because it occasionally sleeps in banana clusters, and is transported to other countries along with the fruit.

PHILANDER OPOSSUM
Caluromys philander

The Philander Opossum is found in woodlands ranging from Venezuela to Brazil. This arboreal marsupial is about 70 cm long. It has a hairless tail longer than its body, which it uses as a fifth limb in climbing. The opossum can hang by its tail when reaching for food, and in this position, it can take eggs from a nest situated on a thin branch out of climbing reach.
As well as birds' eggs, it consumes nestlings and catches insects, smallish tree lizards, frogs and other animals. It also eats plant food, particularly sweet fruit picked from the trees. This opossum seldom descends to the ground to pick up fruit. It forages after nightfall, sheltering by day curled into a ball in a hollow tree.
It is solitary except in the courtship period when it seeks a mate. After a gestation period of 12 — 13 days the female produces about 7 blind, underdeveloped young. These climb into the mother's pouch and attach themselves to her nipples. Here they complete their development during the next two months.

Philander Opossum

Vampire Bat

When they leave the pouch they stay on their mother's back for some weeks, holding her tail with their own. They become mature at the age of 6 — 8 months.

VAMPIRE BAT
Desmodus rotundus

The Vampire Bat is a small bat only about 7 cm long, with a flat, leathery excrescence on its nose. It ranges from Mexico to Paraguay and northern Chile, being found in woodland, especially on the edges of rainforests. It feeds entirely on blood, which it obtains from the bodies of cattle, goats, dogs, poultry and occasionally man. It has 20 teeth. The 2 upper incisors are large and knife-sharp, and the 2 pointed canine teeth are very long and narrow, with sharp cutting edges. The molars are small and are not used. The vampire flies close to the ground after nightfall, looking for its prey. As soon as it discovers a suitable victim it quietly settles on its body and quickly uses its sharp front teeth to cut an incision about 3 — 5 mm deep. Vampires do not suck the blood as is often mistakenly believed, but lap it with their tongues. Their saliva stops the blood from coagulating so the flow is continuous.
Cattle are not much hurt by the actual wounds, but disease and infection may result. These bats transmit various diseases themselves, and the open wounds may become infected with bacteria or parasitic larvae. Insects gathering on the drying blood can transfer any infection. If an animal is regularly attacked by large numbers of bats it will begin to feed irregularly and lose weight. Smaller animals may die due to loss of blood. Although man is not seriously at risk, the use of a protective net at night is recommended in places where vampires are abundant.
The stomach of the Vampire Bat is specially adapted for the consumption of blood, being shaped like a long elastic tube. It can absorb such a large amount of liquid that a replete vampire looks like a ball.
During the day vampires shelter, often in huge numbers, in hollow trees or in caves. They use the same shelters for years or even centuries, and great piles of their droppings accumulate below the places where they roost.
After a gestation period of 105 days, the female gives birth to one, or sometimes two young, which she carries attached to her fur. The young suck their mother's milk until they become independent.

SILVERY MARMOSET
Callithrix argentata

The Silvery Marmoset is one of the best-known of the South American marmosets. It is 45 cm long, its tail measuring 25 cm. It inhabits tropical rainforests of north-eastern Brazil and is particularly widespread in the province of Pará. It frequents the treetops, the lower storeys, and even brushwood. It is also found in wooded savannahs, and often visits plantations or parks on the outskirts of towns. This marmoset lives in family groups of 3 — 6 members, but these may gather into larger troops, which spend the day tracing regular paths through the branches as they search for food. Marmosets take nestlings and are adept at catching small birds, which they kill by biting through the head. The Silvery Marmoset hunts small invertebrates, chiefly insects, but also feeds on vegetable food, and is attracted to plantations by sweet juicy fruit. At night marmosets sleep in hollow trees, often several of them together.
After a gestation period of about 150 days, the female produces twins. She shares parental duties with the male, who often carries the young on his back, returning them to the female to be fed.
The Silvery Marmoset falls victim to birds of prey, arboreal carnivores and large snakes. It is also hunted for its fur.

Common Marmoset

Silvery Marmoset

Black-and-Red or White-lipped Tamarin

COMMON MARMOSET
Callithrix jacchus

The Common Marmoset is distinguished by its long white ear tufts. It is about 60 cm long, including the long bushy tail which can be up to 35 cm long. Except on its thumbs, it has claws instead of nails. It is common in the Amazonian rainforests, living in families or small groups and moving swiftly through the forest canopy. Like most clawed monkeys, it is a clumsy jumper, and young marmosets often fall while leaping from branch to branch, to escape danger. Sometimes they are killed by the fall, and sometimes they become victims of carnivores on the ground. Marmosets have regular pathways through the canopy. Here they hunt insects, take birds' eggs and nestlings and eat sweet fruit. Before sunset they settle in hollow trees to sleep, curling into a ball and using their bushy tails to cover them.
After a gestation period of about 4 months, the female gives birth to twins or occasionally triplets. She carries one infant clinging to her abdomen, while the other rides on her back. Sometimes their father looks after them, carrying them on his back and returning them to their mother for suckling. Young marmosets are independent after 4 months and mature at the age of one year. Like most clawed monkeys, they produce soft, bird-like trilling sounds.
These marmosets are heavily preyed upon by birds of prey and arboreal carnivores. South American Indians hunt them with blowpipes, eating the flesh and making decorations from their skins.

BLACK-AND-RED or WHITE-LIPPED TAMARIN
Saguinus nigricollis

The Black-and-Red or White-lipped Tamarin is a small clawed monkey, reaching a total length of about 60 cm. Interestingly, it has the same number of teeth as man — thirty-two. It lives in forests in the east of tropical South America. During the day, it moves in the treetops, finding clawholds in the finest cracks in the bark. It moves nimbly but jumps clumsily, and prefers not to go down to the ground.

Tamarins associate in families or in small groups. They are diurnal, and settle just before nightfall, in their regular shelters in hollow trees to sleep. A whole group often squeezes into one hollow, each tamarin rolling into a ball and coiling its long, bushy tail around its body.
At sunrise, the group sets out on its regular foraging trip through the treetops. Tamarins comb branches for insects and their larvae, spiders and other invertebrates, occasionally taking birds' eggs or nestlings. They also feed on sweet fruit, green shoots and seeds. They are active throughout the day except in the heat of midday, when they rest in the shade.
After a gestation period of about 145 days, the female normally produces twins. After three days the

Emperor Tamarin

Cotton-head Tamarin

male takes over parental care, carrying the young on his back. Only when they are hungry does he bring them to their mother, who suckles them. Young tamarins reared in captivity become very tame and amusing, but when they are angry, they inflict painful bites. Tamarins mature after 14 months. In the wild they have many enemies among the carnivores and birds of prey, as well as man himself.
Tamarins are able to emit fine bird-like calls.

EMPEROR TAMARIN
Saguinus imperator

The Emperor Tamarin is found in the forests of the upper Amazon. It is about 60 cm long, and gets its name from the long white imperial moustache, which hangs down its chest. The Emperor Tamarin can be found up to a height of 1 000 m in the mountains, but it prefers to live on river banks.
Like other species of tamarins, it is a gregarious monkey, and lives in family groups which wander by day in the treetops. At night, the whole family sleeps in a hollow tree. Here also the female gives birth to twins after a gestation period of 140 — 145 days. The young fend for themselves after 3 — 4 months and are mature at 13 months. The Emperor Tamarin resembles the other species of tamarins in its way of life, but it is predominantly herbivorous.

COTTON-HEAD TAMARIN
Saguinus oedipus

The Cotton-head Tamarin is a small, clawed monkey. It reaches a length of 65 — 70 cm, including its 40 cm-long tail. It inhabits the vast forests of eastern Colombia, and is very abundant in some localities. It lives in family groups composed of the adult pair and their young. Each group occupies its own territory which is defended by the male who loudly warns off intruders. The Cotton-head Tamarin produces a trilling sound, rather like the call of a bird. During the day it moves along regular paths through the trees, climbing nimbly and often making leaps as long as 3 m. It feeds mainly on insects and spiders, but it also catches small lizards and birds, killing them with its small sharp teeth. It also takes birds' eggs. At twilight, the family settles down to sleep in the tree hollow which is its permanent home. There, the female gives birth to twins after a gestation period of 160 — 170 days (some experts say 145 days). She carries them for the first few days, but then the male takes over, carrying them on his back, and returning them to the female only for nursing. At the age of 3 — 4 months, these young tamarins are independent, and become sexually mature after one year. In captivity this handsome monkey lives to be 10 years old. It does not live so long in the wild on account of its many enemies.

GOLDEN LION MARMOSET
Leontideus rosalia

The Golden Lion Marmoset is the largest of the clawed monkeys. It reaches a length of 65 — 75 cm, including a tail 25 — 30 cm long, and its massive mane gives it the appearance of a miniature lion. It is found only in the forests of eastern Brazil, at heights of 500 — 1 000 m. It can no longer be found in some localities, for it has been hunted by the Indians, some of whom keep this monkey as a pet. Its principal enemies are birds of prey and tree carnivores. Lion marmosets live in the treetops where they move easily along branches and up trunks, using their clawed fingers to help them to climb. They have nails only on their thumbs. They live in families composed usually of a pair and its young, but some monkeys are solitary, and some groups are made up of several families. During the day, their soft twittering calls can be heard while they forage for insects, spiders, or the eggs of small birds. They also eat various fruits and shoots. At night, they shelter in dense treetops or in hollow trees, foraging again at dawn. They never voluntarily descend to the ground, but sometimes fall when trying to escape their predators. After a gestation period of 132 — 140 days, the female gives birth to twins, or occasionally to triplets. One infant is carried against her abdomen, and the other on her back. After a few days she often gives the babies to the male, and he takes care of them. The young are suckled for 6 — 10 weeks and then they fend for themselves, although for the most part they stay with their parents.

Golden Lion Marmoset

Common Squirrel Monkey

COMMON SQUIRREL MONKEY
Saimiri sciureus

The Common Squirrel Monkey measures 40 — 75 cm, including its tail which is 15 — 20 cm long. Fully-grown adults vary considerably in size, for reasons so far unknown. Squirrel monkeys are widely distributed in Guyana and in the Amazonas and Pará regions of Brazil. They are resident in forests ranging from coastlands to heights of 600 m in the mountains. They live entirely in the treetops, mainly on the edges of rivers and lakes. During the day, they move carefully among branches, searching for food. They eat various fruits, insects, birds' eggs and tree

Bald Uakari

lizards. They descend to the ground, only to avoid predators such as birds of prey, carnivores or snakes. In the midday heat they rest in the shade of dense branches, where they also spend the night. Squirrel monkeys gather in large troops of over 100, — over 500 have been counted in a single troop, — and members of individual families keep in close contact. After a gestation period of 168 — 172 days, the female produces a single young, which she carries against her abdomen, and which later rides on her back. These monkeys live to be about 12 years old.

BALD UAKARI
Cacajao calvus

The Bald Uakari is found in the rainforests of western Brazil, being particularly common in the area between the Solimões River and the Japurá River — both tributaries of the Amazon. It frequents the tops of large trees, seldom descending to the ground, for the forest floor beneath it is flooded for most of the year. This monkey is about 55 cm long. It moves adroitly among the branches, picking fruit, green shoots and young leaves, or catching insects and small vertebrates, such as young geckos or iguanas. It also often catches small bats in hollow trees. Uakaris live in small groups of one or more families. The males produce harsh calls, and their warning cries carry over a distance of several kilometres. At night, the groups sleep among dense branches, the monkeys huddled together. Nothing is known about the gestation period, or the care of the young, except that the female carries the infant on her back.

Red Uakari

RED UAKARI
Cacajao rubicundus

The Red Uakari is found in wes ern Brazil and eastern Peru in an area bounded in the north by the Solimões River, in the west by the Ucayali River, and in the east by the Juruá River. It is about 60 cm long and has a stumped tail only 8 cm long. This monkey lives in the treetops, usually along river banks. Its way of life does not largely differ from that of the Bald Uakari.

DOUROUCOULI
Aotus trivirgatus

The Douroucouli is widely distributed from Guyana across the Amazon region to Peru. It is 85 cm long, of which 50 cm is taken up by the tail. This monkey is characterized by its large eyes, a sign that it is nocturnal, which is unusual in monkeys. Its sense of hearing is very acute. During the day, groups of douroucoulis sleep in hollow trees. After nightfall, the monkeys set out to forage for insects and small lizards, birds and mammals. They also feed on sweet fruit. They move nimbly in the treetops, leaping from tree to tree.

The gestation period is estimated at 140 days, and twins are quite often born. At first the mother carries her baby clinging to her chest, but after a few days it moves on to her back. After about 9 days, the infant is carried by its father. When it is hungry, the baby makes a twittering sound which encourages the father to take it to its mother to be suckled.

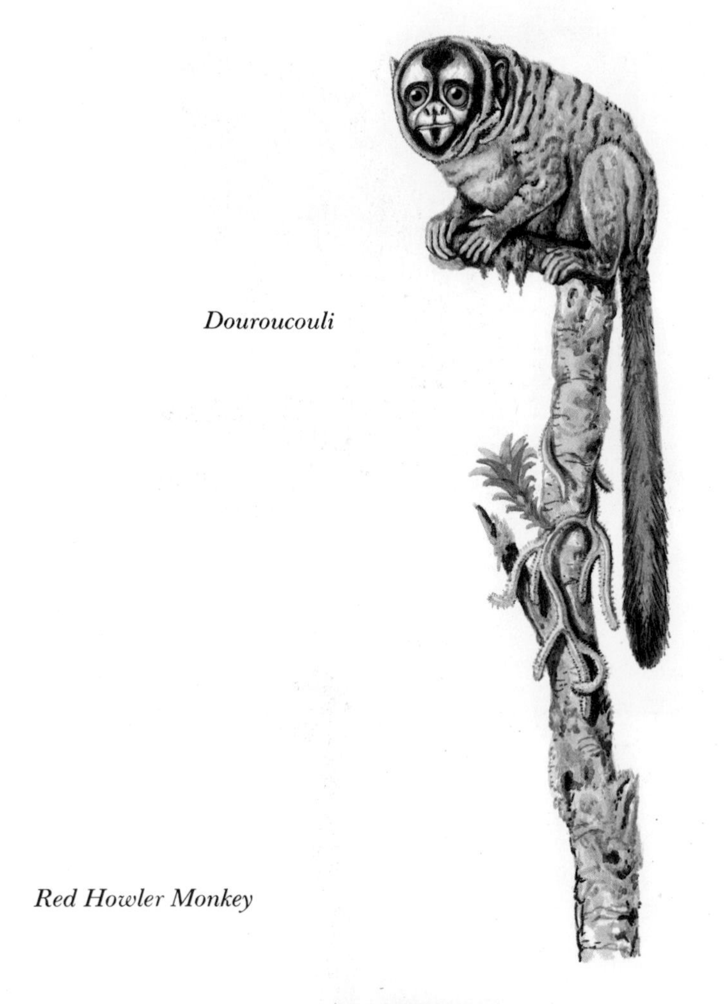

Douroucouli

RED HOWLER MONKEY
Alouatta seniculus

The Red Howler Monkey is one of the best-known monkeys of the extensive primary forests of the central Amazon region and Colombia. It is up to 135 cm long, including the tail which measures about 70 cm. The end third of the tail is naked on the underside. It is strongly prehensile, being used to hang and swing on branches. Females are much smaller and more delicately built than males. The Red Howler Monkey is a typical inhabitant of the canopy. It climbs adroitly and leaps great distances from tree to tree. It comes down from the trees only when forced by a predator, for it moves clumsily on the ground. If it accidentally falls into the water, it is able to swim, but it does not enter the water from choice.

Howler monkeys usually live in troops of about 20, but there may be as many as 50 in some. Each troop is usually made up of 3 males and about 8 females with their young. The males share the leadership without hostility, and defend the territory and announce its occupation by loud howling. This noise is made by means of a special vibrating drum in the throat which intensifies the voice.

Howler monkeys are strictly vegetarian, feeding on

Red Howler Monkey

Brown Capuchin

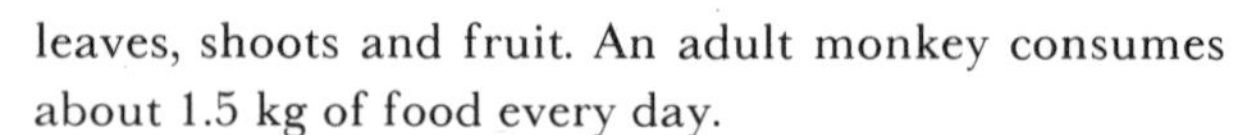

leaves, shoots and fruit. An adult monkey consumes about 1.5 kg of food every day.
After a gestation period of about 140 days, the female gives birth to a single young which she carries on her back. The young are dependant on parental care sometimes for more than 18 months. They become mature at the age of about 4 years. Their principal enemies are large birds of prey, particularly harpy eagles, and large, climbing carnivores. Howler monkeys are hunted by man for their meat and for their thick, high-quality fur.

BROWN CAPUCHIN
Cebus apella

The Brown Capuchin, like the other species of capuchin monkey, is common in the tropical forests of South America. It has a characteristic 'cowl' of thick hair on the crown of its head, which gives rise to its name and it has a long prehensile tail, used in climbing. When walking, the capuchin carries its tail coiled at the tip. Capuchins live in groups of up to about 40, led by a strong male. Although respected by the other members, the dominant male has to fight for leadership, and duels often end in the death of one of the protagonists. Capuchins have extremely powerful teeth which can cause serious wounds.
Capuchins seek their food among the trees, and on the ground where they collect fallen fruit or nuts and nibble at green shoots. They also eat insects and take birds' eggs and nestlings. They catch and collect their prey skilfully with their hands and can manipulate primitive tools, such as stones, which they use for cracking nutshells. They place tough nuts on top of large flat stones, and keep hitting them with another stone held in the hands until the shells are broken. Capuchins are surprisingly strong and often use stones weighing as much as they themselves do.

Black Spider Monkey

At night capuchins shelter in hollow trees or among dense branches, but they often fall victim to large birds of prey, for their strong teeth are no match for the birds' formidable claws. After a gestation period lasting 180 days, a single young, or exceptionally twins, is born. It is carried on the mother's back and is suckled for up to 5 months.

BLACK SPIDER MONKEY
Ateles paniscus

The Black Spider Monkey has extremely long limbs and a conspicuously small head. It measures 130 cm, but the tail takes up over 70 cm of this. It is common in the tropical primary forests of South America, where it moves dexterously in the canopy, gripping the branches with its prehensile tail. It uses its tail to hang from branches, to pick fruit, and to pass food to its mouth. This agile monkey runs along the branches on its hind legs, holding its tail up like a question mark. It can swing by one arm and then make a flying jump as far as 10 m, before grasping a branch with the other hand. It seldom visits the ground, for there it moves slowly and clumsily. Before nightfall, groups of spider monkeys gather in the branches, sitting close to each other for warmth.
They feed mainly on fruit, shoots and leaves, but they sometimes catch insects or take an occasional nestling, egg or small iguana.
After a gestation period of about 140 days, the female gives birth to a single young, which she carries first on her chest, then on her back. The infant clutches its mother's hair and coils its tail around hers. Spider monkeys have a relatively long lifespan of up to 20 years.

Humboldt's Woolly Monkey

HUMBOLDT'S WOOLLY MONKEY
Lagothrix lagotricha

Humboldt's Woolly Monkey is the largest South American monkey. It reaches a length of 1.5 m, which includes its bushy, prehensile tail which is 70 cm long. It is a slow, hesitant monkey with long, thick, soft fur to protect it from the cold, for it lives in mountain areas up to a height of 3 000 m.
This is a sociable monkey. It forms small troops of 12 — 15, often in company with howlers and capuchins, and only old males are sometimes solitary. They wander through the canopy, remaining always within their own territory. The leading male announces the presence of the troop in a sonorous howling voice, warning other troops to keep away. However, the dominant male is not ferocious and never provokes fights with intruders, preferring to intimidate them with his facial expressions.
On their foraging trips among the trees, woolly monkeys feed chiefly on shoots, leaves, fruits and nuts, though they occasionally take birds' eggs or catch small animals. When climbing, they use their prehensile tails in the same way as other South American monkeys. At night, they sleep with their tails coiled around the branches.
A single young is born after a gestation period last-

Kinkajou

Red Coatimundi

ing about 140 days (up to 225 days according to some observers). It is carried on its mother's back with its tail coiled around hers. The young monkey is suckled for more than a year, although it also takes solid food after the first few months. Humboldt's Woolly Monkey matures after 4 years and lives to be about 20 years old.

Among the enemies of this monkey are large birds of prey, particularly harpy eagles, and man. It is hunted by native Indians for its meat and fine fur, with the results that it is no longer found in the neighbourhood of villages.

KINKAJOU
Potos flavus

The Kinkajou is a small bear-like carnivore. It inhabits the forests of South and Central America, and is variable in colour. Its total length is about 95 cm, about half of which is made up of the tail. Kinkajous push their elongated snouts into cavities and cracks, seeking prey both by touch and smell. They are excellent climbers, making use of their long, prehensile tails as they search for the nests of wild bees. They are attracted by the scent of honey which is found in large hollows in which bees build their colonies. They tear up the honeycombs with massive claws on their short but robust feet, and lick the honey with their long tongues. Their thick fur protects them from the angry bees. Kinkajous also take birds' eggs and nestlings, and they eat insects and molluscs. The bulk of their food consists of fruit, taken from plantations near the edges of the forests.

After a gestation period lasting 112 — 118 days, the female gives birth to 1 — 2 young in a tree hollow or small cave. At birth the young weigh about 150 g and are blind, opening their eyes after 7 — 14 days. They are suckled for 16 — 18 weeks, but begin to take solid food after 7 weeks. At the age of 3 — 4 weeks, they leave their shelter for the first time, and play nearby. They are mature at the age of 18 months.

Native Indians often catch young kinkajous and keep them as pets.

RED COATIMUNDI
Nasua nasua

The Red Coatimundi is distributed in the tropical forests of South and Central America. It has also been found in the south of the United States. It is a bear-like carnivore, over 1 m long, with a pronouncedly long snout. It is agile both in the trees and on the ground. When running, it holds its bushy tail erect. It pushes its way through the undergrowth, into caves, among rocks and in the treetops, thrusting its sensitive snout into crevices and holes. It has an excellent sense of smell, and touch cells at the tip of the snout help it to detect its prey. It feeds on insects, molluscs and other invertebrates, and takes birds' eggs and smallish young mammals. It also eats plant material, favouring fallen fruit. During its nightly foraging trips it sometimes visits henhouses and kills poultry, which makes it unpopular with the native population.
Coatimundis usually live in troops composed of several females and their young, or in families of one female and her offspring. The males are solitary for most of the year, seeking partners only in the breeding season. After a 71 — 74-day gestation period, the female produces a litter of 2 — 7 young. At first they have little or no fur and are unable to see. Their ear orifices open after 4 days and their eyes after 11 days. They suck their mother's milk for 24 — 26 days, but are able to take solid food by the age of 19 days. They are very playful and leave their den when they are 3 — 4 weeks old to romp in front of the entrance. Forty days after birth they forage for themselves, but stay with the family. They become mature at the age of 2 years.

TIGER CAT
Felis tigrina

The home of the Tiger Cat is in the vast forests ranging from Mexico to Argentina. This small cat is about 80 cm long, including the tail, and has variable coloration.
It lives mainly in the tops of the trees where it hunts birds, small mammals and iguanas. It climbs dexterously and moves with agility along the branches. It is a good swimmer when forced to enter the water. In the courtship period the male seeks a partner, but becomes solitary again after mating. After a gestation period of 74 — 76 days, the female produces one or two kittens in a tree hollow. The young gain their sight when they are 10 days old. They leave the hollow for the first time after about 4 weeks, but

Tiger Cat

Ocelot

return to it regularly. They suck their mother's milk for 6 — 8 weeks.
The Tiger Cat often falls victim to large birds of prey and boa constrictors.

OCELOT
Felis pardalis

The Ocelot is a large wild cat. It reaches a total length of 95 — 135 cm, a third of this being taken up by the tail. It is about 50 cm high at the shoulders. Coloration and pattern vary from individual to individual. The Ocelot is distributed from South America, home of the largest specimens, across Central America to Texas in the United States. This handsomely marked cat is resident in thick woodland. It is an expert climber and hunts birds and small arboreal mammals, as well as preying on rodents and crustaceans on the river banks. It is adept at swimming and is often seen in water, easily crossing wide rivers — providing it escapes the crocodiles. Like many cats, ocelots are solitary, seeking partners only in the breeding season. The litter of 1 — 4 young is born in a safe shelter, usually a tree hollow, after a 70-day gestation period. The newborn kittens are blind and open their eyes after 7 – 10 days. They leave the den when they are 3 — 4 weeks old and play nearby. After 50 — 55 days, young ocelots can digest meat brought to them by their mother, and after 2 months they are able to catch their own prey.

JAGUAR or AMERICAN TIGER
Panthera onca

The Jaguar or American Tiger is the largest species of cat on the American continent. It is relatively widespread, ranging from South to Central America, and up to southern Arizona in the United States. It was at one time found in southern California and southern Texas, but it has now been exterminated there. Jaguars are 165 — 240 cm long, including the tail. Curiously, jaguars occurring in southern regions are much larger than those found further north. The largest specimens, weighing over 130 kg, are found in the state of Mato Grosso in Brazil. Black individuals (melanistic forms) may occasionally appear. Melanism is an inherited characteristic, although black individuals may be born to normally coloured parents.

The Jaguar is not restricted to forests but also frequents river banks with reed beds and tall grass. Here it can hide easily in the thick cover and crawl unseen towards its prey. For the most part it seeks capybaras or young tapirs, but when food is in short supply it also feeds on small mammals and birds, and on fishes caught in shallow water. An expert tree-climber, it often attacks monkeys. Near villages it is a threat to domestic animals, and it sometimes attacks even cows and horses. Adult jaguars have virtually no enemies in the wild, unless they enter water. They are good swimmers and can easily cross wide streams, but while in the water they may fall prey to both crocodiles and man. Native hunters can get close to a jaguar by boat, to spear it — but there is always a risk that the jaguar will get into the boat!

For most of the year, jaguars are solitary, each occupying its own limited territory, but males cease to respect territorial rights in the courtship period when they set out to find a mate. Then fierce fights take place when two rivals confront each other. The male stays with his mate for only a few days and then leaves her. After a gestation period of 93 — 110 days, the female gives birth to 1 — 5 young in a thicket or small cave. They weigh 500 — 900 g at birth and are blind, opening their eyes after 10 — 13 days. At the age of 40 days, they leave the den for the first time, and after 45 — 50 days, they begin to eat meat brought to them by their mother. Young jaguars are suckled for 5 — 6 months, and stay with their mother for about 2 years. Females mature after 2 — 2.5 years, males after 3.5 years. They become fully grown after 5 — 6 years.

Jaguar or American Tiger

Three-toed Sloth

THREE-TOED SLOTH
Bradypus tridactylus

The Three-toed Sloth is a very curious creature which inhabits tropical regions of South and Central America. It is 50 cm long and is unusually adapted for life in the treetops. The sloth has 9 neck vertebrae and can turn its head through an angle of 270°. Although it belongs to the order Edentata (toothless mammals), it has 18 teeth, 10 in the upper jaw and 8 in the lower. These teeth are composed only of dentine, and have no enamel coating. Its stomach consists of four parts, one of which contains bacteria and protozoans that aid digestion of tough leaves. Sloths feed mainly on leaves, but they sometimes eat soft fruits which have no stones in them. They cannot digest stones or pass them unchanged through their digestive system, so if any enter the stomach they accumulate and may cause death. The digestive system is such that the sloth only needs to defecate once every 10 days or so. Faeces accumulate in the anal orifice, where a parasitic scarab beetle is often found. Body temperature is another peculiarity of this animal, for its temperature varies between 27 and 35° C, frequently falling below that normal for mammals.

The Three-toed Sloth has three toes on each of its front and hind limbs. These toes bear strong hooked claws on which the animal can hang. Throughout its life the sloth moves upside down. Its hairs point downwards when it is hanging underneath a branch, so in the rainy season the water easily drips off. In the warm atmosphere of the humid tropical rainforests, a species of green algae grows on the sloth's hair making it a greenish colour, and providing it with valuable protective coloration. In the treetops, this defenceless animal becomes almost invisible as a result of its coating of green algae.

The sloth moves unhurriedly when climbing, but it is capable of achieving a speed of up to 8 km per hour. It descends to the ground only rarely and when there it can only crawl or drag itself along with difficulty. If it accidentally falls into water, it swims vigorously, but it does not enter water willingly.

During the day the sloth usually sleeps squeezed in a fork between two branches, where it can rest the muscles of its limbs. It sets out to forage at twilight. It has poor vision, but this is compensated for by an excellent sense of smell. The sense of hearing is important for a nocturnal animal and helps to locate a mate in the breeding season. Sloths produce high-pitched wailing noises. For most of the year, they are solitary, the partners separating after the breeding season.

Following a gestation period of about 170 days, the female produces a single young which stays with its mother for almost a year. The infant firmly clutches its mother's abdominal hair with its claws, its head towards hers. It is born covered with hair and well developed. At the age of 14 days, it begins to nibble young leaves, and is weaned on to solid food by the age of 10 weeks. A young sloth becomes entirely independent at 9 months of age and sexually mature when it is about two and a half years old.

TAMANDUA
Tamandua tetradactyla

The Tamandua is distributed mainly in tropical regions of South America, its range extending as far north as southern Mexico and south to northern Argentina. In some places, it is very abundant. It

Tamandua

reaches a length of 100 cm, but almost half of this is a thin prehensile tail, used for grasping branches when the animal is climbing. The Tamandua spends most of its life in thick treetops, looking for nests of tree termites. It breaks up the spherical structures with its strong claws and picks out the termites with its tongue. It seldom descends to the ground, and moves only slowly there. It is a good swimmer. In the courtship period the partners stay together for only a short time. After a gestation period of about 190 days, the female produces a single young which she carries about with her, looking after it for almost a year.

GIANT ARMADILLO
Priodontes giganteus

The Giant Armadillo is a real armoured monster of the South American tropics. It is 175 cm long, including its 75 cm long tail, and it weighs up to 60 kg. It frequents thinly forested areas and bushland in Paraguay, but it is relatively rare. It has huge, spade-like claws which enable it to dig quickly in soft soil. It excavates tunnels up to 5 m long and 1.5 m deep, which lead to a lair with several escape routes. The armadillo sleeps by day, setting out after sunset to find food. It combs the neighbourhood around its

Giant Armadillo

Prehensile-tailed Porcupine

lair collecting fallen fruit, taking birds' eggs and nestlings, and catching small mammals, insects, molluscs and worms. It is predominantly carnivorous.
After a gestation period of four months, the female produces one or two young in the lair and suckles them for a month, after which time the young fend for themselves.

PREHENSILE-TAILED PORCUPINE
Coendou prehensilis

The Prehensile-tailed Porcupine is a rodent about 60 cm long. It is distributed in Venezuela, Guyana, the adjacent part of Brazil and the island of Trinidad. It has many short quills among its hair, protecting it from predators. This arboreal rodent climbs skilfully but slowly, grasping the branches with its long, thin prehensile tail. It feeds on shoots, leaves, blossoms, fruits and tree bark. It occasionally comes down from the trees to gnaw at roots. It is nocturnal and leaves its shelter after sunset, sleeping by day in tree hollows or forks of dense branches.
The Prehensile-tailed Porcupine is solitary, looking for a mate only in the breeding season. After a gestation period of 210 — 217 days, the female gives birth to a single young, born with open eyes and sparse soft quills. The young porcupine is well developed and is able to climb when it is only two days old. Sexual maturity is reached at 2.5 years and the lifespan averages 9 years.

PACA
Cuniculus paca

The Paca is the largest South American rodent after the Capybara. It reaches a length of 70 cm and a weight of 9 kg, females being usually smaller than males. The Paca inhabits the extensive woodland and bushland from Argentina to southern Mexico. It lives on the edges of forests or in thickets, usually near water. It digs burrows 1 — 2 m long in which to shelter during the day. It is nocturnal, feeding after twilight on grass, shoots and fallen fruit. Pacas in large numbers cause considerable damage in maize fields. They also ruin other crops and so the local peasants exterminate them if they can. Pacas are also hunted for their meat, which is considered to be a delicacy.
After a gestation period of about 105 days, the female produces one, or occasionally two young, already covered with hair, and able to see. She usually has two litters each year.

CUBAN HUTIA
Capromys pilorides

The Cuban Hutia is a tree rodent which lives in forested areas of Cuba where it is a protected species. Its length is 45 — 60 cm, including a tail 15 cm long, and it weighs up to 8 kg. The hutia is a good climber, grasping branches with its prehensile tail. It is predominantly nocturnal, feeding after sunset on young twigs, leaves, blossoms, fruit and seeds. It can some-

Paca

times be seen during the day, basking in the morning sun, but most of the time it hides in tree hollows, rocky crevices or underground holes.
Hutias live in pairs or in families. After a gestation period of 120 — 130 days, the female produces 3 — 4 young, able to see and immediately very active. They can move around independently a few hours after birth. After 10 days, they eat young shoots and leaves, although they continue to suck their mother's milk for 6 weeks. The average lifespan of the hutia is 10 years.
Other species of this tree rodent were recently discovered in Cuba.

Cuban Hutia

COLLARED PECCARY
Tayassu tajacu

The Collared Peccary is distributed throughout Central America, and is also found as far south as Argentina and as far north as southern Arizona and southwestern Texas in the United States. In this extensive territory there are 14 subspecies, differing in size and

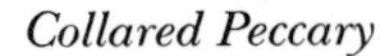

Collared Peccary

coloration. It is about 95 cm long, 40 cm tall and weighs up to 25 kg. The peccary inhabits thickets along the edges of forests, and areas of brushwood. It also lives in swamps on river banks, easily swimming across the streams. It forms small herds of about 12 individuals which wander through thickets, grazing, grubbing up roots, finding fallen fruits, and also taking some animal food. Peccaries often catch reptiles, including venomous snakes. They are protected from the snake's fangs by thick hide covered with bristly hair, and by a layer of subcutaneous fat. They also eat the eggs of ground-nesting birds, and feed on insects, molluscs and worms. On river banks, they sometimes find dead fishes or hunt crustaceans. The herds always run as fast as possible across open space to seek shelter in brushwood, for they have many enemies. Peccaries rarely escape large carnivores, such as jaguars or mountain lions, but they often confront smaller carnivores, such as wild dogs. They defend themselves with their powerful canine teeth and cause serious wounds. They often succeed in chasing the attacker away. Young peccaries fall prey to large snakes like anacondas, and to large birds of prey. In their native land, peccaries are the favourite game of native hunters, who use dogs to track them and to flush them from their refuges. Peccaries have even been introduced in places where they do not occur naturally, for hunting purposes. Breeding can take place in any season. The female usually gives birth to twins in thick cover, often on a swampy island. A single young is rare and a higher number is exceptional. The gestation period lasts 142 — 149 days. At first the nursing mother and her young stay apart, but within 7 — 10 days young peccaries are able to run and can join the rest of the herd. Suckling is very brief and the young fend for themselves after 6 — 8 weeks. Females mature at the age of 9 months and males at one year of age.

BROCKET DEER
Mazama americana

The Brocket Deer is only 60 cm tall at the shoulders and weighs 10 kg. It has tiny, straight antlers, 5 cm long. Its home is in the dense South American forests ranging from Guyana to Paraguay. It is very common in some places but is rarely seen because it

Brocket Deer

hides by day in impenetrable thickets and grazes or browses at dusk. It is usually solitary, forming pairs only in the breeding season, which can be at any time of the year.
At night, this small deer ventures across open spaces along the edges of forests and sometimes visits nearby plantations. It is very shy and wary, yet it often falls prey to carnivores, large birds of prey and snakes. Young, inexperienced fawns are particularly at risk. A boa constrictor, lurking motionless in a bush, can pounce on this small deer, seize it with its sharp teeth and crush the victim in its deadly coils. The animal dies almost immediately and the boa swallows it whole.
After a gestation period of 185 — 210 days, the female gives birth to a single young. For the first few days, the fawn lies concealed in tall grass and the mother comes to suckle it. Young brocket deer are capable of walking and running from the age of two days, but only follow their mother after 14 days. Females mature after one year, males after 18 months.

Muscovy Duck

MUSCOVY DUCK
Cairina moschata

The Muscovy Duck is common throughout its range, from the forests of central Mexico south to northern Argentina and Uruguay. Both sexes are identically coloured, but the male weighs 3 — 5 kg and has a prominent protuberance above his beak, while the female weighs only 1.6 — 2.8 kg and has a smaller protuberance. This duck is found near lakes and rivers. It roosts on the branches of tall trees where it is protected from enemies on the ground, but it often falls victim to birds of prey, and the local Indians hunt it for its palatable meat.
The breeding season of the Muscovy Duck coincides with the beginning of the rainy season, which means March in Peru and Mexico, and November in Bolivia. During this period the males engage in frequent disputes, for the pair bonds are not strong and a male usually jealously watches several females. The female nests practically anywhere: in tree hollows just above the ground or as much as 20 m above it; in the abandoned nests of birds of prey; in palm trees; and occasionally on the ground, in brushwood or in rocky crevices. A clutch of 10 — 15 whitish or greenish-white eggs is laid and covered in a thick layer of down. The female incubates the clutch for about 34 days and cares for her brood while the male defends the territory and chases away intruders regardless of their size.
The Muscovy Duck feeds on green plants, seeds and fruit, as well as insects, worms, molluscs and other small invertebrates.
This species is bred as a domestic duck all over the world.

SOUTH AMERICAN KING VULTURE
Sarcorhamphus papa

The South American King Vulture is a heavily built, beautifully coloured bird of prey. It is about 85 cm long and has a wingspan of up to 2 m. It inhabits tropical regions of South America from central Mexico to northern Argentina. It also occurs in Trinidad and has been seen in Florida, although it does not nest in these areas. It lives in forests or wooded savannahs, mainly in the lowlands, but sometimes in the mountains up to a height of 1 000 — 1 300 m. Although it is common in some areas, little is known about its life in the wild. King vultures live in pairs or in families, the young staying with their parents as long as two years. Adult birds probably breed only once every two years. Young birds can be distinguished by their coloration, for up to the age of 3 years they are black above and whitish below.
This vulture builds its nest among the crowns of tall

South American King Vulture

trees, usually more than 20 m above the ground, or sometimes on protruding rocks. The female lays a single off-white egg with russet spots, and both partners share in incubation for 56 — 58 days. When the chick hatches, both parents feed it on food regurgitated from their crops. Adult birds usually feed on the carcasses of large game, but they have been seen to hunt various small reptiles.

HOOK-BILLED KITE
Chondrohierax uncinatus

The Hook-billed Kite is found in tropical and subtropical America in an area ranging from northern Mexico to northern Argentina. It also occurs in the West Indies and has been seen in Texas. It is a small bird of prey, the female weighing about 320 g and the male 250 g. The chief characteristic of this bird is the bill, which has an unusually long and hooked upper part. The female is brownish and has a grey head and russet underparts with white crossways stripes. The male is greyish. In the black (melanistic) form, both sexes are alike.
The Hook-billed Kite is a forest dweller, preferring the lowlands but sometimes occurring up to a height of 2 000 m. It leaves the primary forests to fly to the coffee plantations where it perches on isolated tall trees from where it hunts for molluscs, which are its staple diet. In Cuba, it catches mainly tree molluscs of the genus *Polymita,* but it also feeds on terrestrial species. The long upper mandible enables the bird to extract the soft bodies from their shells. Kites occasionally also catch insects, collect larvae, and eat small tree frogs.
This bird of prey usually lives in pairs, and families stay together after fledging, unless the young birds become independent particularly early. The pairs build nests in tall trees, collecting twigs in their bills. The female lays 2 whitish eggs with chocolate-coloured speckles. So far, little is known of the nesting habits of this interesting bird despite its common occurrence in some localities.

PLUMBEOUS KITE
Ictinia plumbea

The Plumbeous Kite is found in the tropical zone of Central and South America, ranging from central Mexico to northern Argentina and Paraguay. It is about 32 cm long, and has a wingspan of 95 cm. It is resident mainly in mountain forests at a height of 1 000 — 1 300 m, but is also found in mangrove swamps along the coast. In autumn, these kites fly from Mexico and Central America to tropical South America, and the birds living in colder regions of Argentina migrate northwards. The migrating birds merge into huge flocks, sometimes of several thousands, giving rise to formations over 1 km long! They often join other species of kites, and a gigantic flock of these beautiful raptors is a magnificent spectacle.
Their diet consists mainly of insects, skilfully caught

in the claws in flight. When termites are swarming, scores of kites gather to catch them. They are also adept at hunting wasps and bees. Molluscs and small birds or iguanas complete their diet.
On the coast the nest is built 10 — 12 m above the water at the end of a mangrove branch. A forest nest is situated in a treetop 30 m above the ground. It is made of twigs and lined with shoots and leaves. The female lays 2 — 3 off-white eggs with brown spots. Both partners incubate the clutch for 30 days and feed their young on insects every 10 minutes, putting them directly into their bills. Older chicks pick up the prey from the nest. After 36 days the young leave the nest and learn to hunt.
The Plumbeous Kite is an expert flyer, and capable of deterring other birds of prey much stronger than itself.

Hook-billed Kite

BLACK-CHESTED BUZZARD EAGLE
Geranoaetus melanoleucus

The Black-chested Buzzard Eagle is a relative of the kites. It inhabits predominantly wooded mountain localities and rocky sites, though in Peru it is also common in hilly woodlands near the coast. It ranges from Venezuela in the north to Argentina in the south. This bird of prey is about 70 cm long and has a wingspan of up to 180 cm. Both sexes are alike in coloration. The female is often bigger by one third, weighing up to 3 kg. The chief haunt of these birds is on the branches of tall trees near the edges of forests or on high rocks affording a good view of their hunting grounds. They sometimes fly just above the ground, looking for prey, especially small mammals. They catch young skunks and insects, particularly beetles, and occasionally peck at juicy fruit.
The nest is usually built on a rock, often above a river, at a height of 300 m. It may also be situated on the branch of a tall tree, and the same nest may be used for several years. Occasionally a nest may be built in a bush or on the ground.
The female lays 2 white eggs, usually with olive-brown spots. Both partners share in incubation for 30 days, but the female spends most time on the nest. At first the young are fed only by the female, who

Plumbeous Kite

Black-chested Buzzard Eagle

receives food from the male, but later both parents feed the chicks. The young birds leave the nest when they are about 50 days old, but they are fed for a further 3 — 4 weeks while they are being taught to hunt.

HARPY EAGLE
Harpia harpyja

The Harpy Eagle is a massive bird of prey with extremely powerful feet and huge claws. The male weighs up to 4.5 kg, measures about 100 cm in length, and has a wingspan of more than 250 cm. The female is sometimes even bigger. This is the strongest raptor in the world. It is distributed in the tropics of Central and South America, where it inhabits lowland forests and wooded hills from the south of the Mexican state of Vera Cruz to eastern Bolivia. It is an expert hunter of all species of monkeys, even in the dense treetops. As soon as this fearsome raptor appears, monkeys flee in panic, but harpy eagles easily catch up with an exhausted capuchin or howler monkey and grab it with their long, pointed claws. They also catch coatimundis and large birds, like macaws and toucans, and hunt agoutis, snakes and lizards.
The nest is built among the treetops, often 60 m above the ground. It is made of strong branches and is about 120 cm tall and 150 cm across. The female lays 1 — 2 white, yellow-spotted eggs in the nesting depression, which is lined with green leaves. Although this raptor is quite common, little is known of its nesting habits. Ornithologists believe it breeds only once every two years. The young are dependent on parental care up to the age of 5 months when they are capable of flight.

ORNATE HAWK-EAGLE
Spizaëtus ornatus

The Ornate Hawk-eagle is distributed from central Mexico to northern Argentina and Paraguay. It also lives on the islands of Tobago and Trinidad, and is quite common in some localities. It is found in the vast tropical forests up to a height of 3 000 m, but it also visits open countryside, bushland savannahs and plantations. It hunts fowl-like birds, herons, parrots

Harpy Eagle

Ornate Hawk-eagle

Isidor's Eagle

or small mammals. It may even attack condors. The male is 50 cm long, weighs about 1 kg and has a wingspan of about 130 cm. The female weighs up to 1.5 kg. The sexes are alike in coloration but the plumage around the heads of young birds is predominantly white.

The Ornate Hawk-eagle lives in pairs throughout the year. It builds a large nest, 150 cm across, in treetops up to 30 m above the ground. The nesting depression is lined with green leaves. Although this is a very common raptor, little is known of its nesting habits.

ISIDOR'S EAGLE
Oroaetus isidori

Isidor's Eagle is a sturdy bird of prey, 70 cm in length and having a wingspan of 180 cm. The female

is slightly larger than the male. This beautiful eagle inhabits wooded slopes of the Andes in the subtropical zone, at a height of 1 700 — 2 700 m and it sometimes visits alpine locations up to 3 300 m. It is found in the region from Venezuela to Bolivia and north-western Argentina. It feeds mainly on small arboreal mammals including monkeys, but it also hunts birds and iguanas. It makes its way through the treetops with great ease and its prey has little hope of escape.
Isidor's Eagle builds a large nest, almost 2 m across and 90 cm tall, in treetops 20 — 30 m above the ground. The nest is made of dry branches and is often used for several years. In Colombia where this raptor is abundant, the nests are built in oak of the species *Quercus colombiana.* The nest is lined with green twigs which the birds tear off with their bills. They repair the nest continually throughout the nesting period. The female usually lays a single white egg with chocolate-coloured specks, and incubates it for 36 days while the male feeds her. He later brings food for the chick and passes it to his mate, who tears it up into pieces. The young eagle is a light colour, quite unlike its parents. It is capable of flight at the age of 4 months.

Laughing Falcon

LAUGHING FALCON
Herpetotheres cachinnans

The Laughing Falcon is widespread in the tropics from Mexico to northern Argentina and Paraguay. Its wingspan is about 95 cm, and the raptor itself is 40 cm long. It is confined to damp situations in sparsely wooded lowlands, and to savannahs with isolated trees. It usually lives in pairs, but outside the breeding season it may briefly become solitary. Falcons choose sites on branches, from which they set out to hunt and to which they return with their prey. Snakes, which are the mainstay of their diet, are grasped by the head and killed. Smaller snakes are swallowed whole, tail first, but larger prey is torn into pieces. The Laughing Falcon catches both venomous coral snakes and harmless tree snakes, and occasionally eats small rodents and iguanas.
The Laughing Falcon seldom builds a new nest but takes over the deserted nest of some other bird of prey. A nesting pair may settle in a hole in a tree more than 30 m above the ground, or choose a rocky crevice. The female lays a single egg, pale carmine in colour with many dark brown specks. She incubates it for about 30 days while the male brings food to her. He also feeds the newly-hatched chick though later both partners feed it. After about 40 days the young falcon leaves the nest but remains dependent on its parents for a further 25 days. It is fed until it learns to hunt.

OCELLATED TURKEY
Agriocharis ocellata

The Ocellated Turkey is a bird of Central America. It is widely distributed particularly in Guatemala and Honduras, but it is also found in south-eastern Mexico, in the Yucatan region. It is about 75 cm long, the female being much smaller and less conspicuously coloured.
The Ocellated Turkey is resident mainly in light woodland, open bush, and sometimes in fields or

Ocellated Turkey

plantations where it finds its food. It feeds on insects, spiders and invertebrates, and often catches small reptiles and young rodents. It also eats seeds, berries, fallen fruit and green shoots.

In the courtship period, the males set about attracting the attention of the females. They find open spaces and engage in nuptial displays, stretching out their colourful tail feathers, and running in circles with their wings hanging low. The coloured outgrowths on their heads and necks swell at this time. The Ocellated Turkey is polygamous, each male having a small flock of hens. He takes no part in bringing up his offspring. The female prepares the nesting depression under a dense bush among stones or behind a fallen tree trunk. She lines it with dry leaves and grass and lays 8 — 15 eggs. Since the clutch is on the ground and not in the safety of a tall treetop, it is an easy prey for many animals. For this reason, these fowl-like birds lay a large number of mottled eggs, which merge with the background. The eggs of this species are pale yellow to ochre in colour, with a dense pattern of dots and russet or brown spots. The female incubates the clutch for 28 days. The newly-hatched young become dry on the second day after hatching and leave the nest. Their mother accompanies them and at night covers them with her wings to warm and protect them. When they are 10 days old, the chicks begin to try to fly, and as soon as they are successful, they roost with their mother and the other adult birds on branches. When the young leave the nest, the turkeys gather together in flocks.

GREAT CURASSOW
Crax rubra

The Great Curassow is a robust fowl-like bird, which can attain a length of 90 cm. The female is slightly smaller and has a black, white-spotted crest. This bird ranges from Mexico to western Ecuador and inhabits the extensive primary forests. It spends most of the day in the trees, looking for food. It pecks green shoots, fruits, seeds, and catches insects and invertebrates. It occasionally eats a small iguana or gecko, and eats fallen fruit. It is active by day, at dusk flying to the tallest branches and spending the night there.

The nest is also built in the trees. It is flat, woven from twigs and leaves, and situated high above the ground. Curassows often use the abandoned nests of birds of prey to form the foundations of their own nests. The female lays only two eggs, dull white and the size of goose eggs. She incubates them for 25 —

Great Curassow

Nocturnal Curassow

29 days. The hatched young have to jump down from the nest, which may be 6 — 30 m above the ground, because they are not yet able to fly. At first the hen and her chicks hide in thickets. The young begin to fly after 4 days, and after a week they can fly up to low branches to roost. By the time they are a month old, they can fend for themselves.
They are preyed upon chiefly by large raptors, but they are also hunted for their palatable meat. Indians sometimes put curassow eggs under domestic hens for incubation. The chicks are easily tamed and stay with the barnyard fowls.

NOCTURNAL CURASSOW
Nocthocrax urumutum

The Nocturnal Curassow is a denizen of the dense primary forests in the basin of the Rio Negro in northern Brazil, and in adjacent parts of Colombia and Venezuela. It is about 80 cm long and both sexes look alike. This fowl-like bird is active mainly at night, especially in moonlight. It feeds on insects, larvae, worms, molluscs, and small vertebrates, such as geckos. It also eats fruit, particularly berries. In dense, dark undergrowth, it also forages early in the morning and before twilight. During the day, it usually hides among the branches.
It builds its nest in the daytime, on branches or in the deserted nests of other birds. It is a small simple structure, consisting only of a flat heap of twigs and leaves. At night in the courtship season, the male produces hollow sounds like those of a distant trumpet. The female lays 2 whitish eggs, incubates them for about 29 days, and cares for the young. After a week, the chicks can fly a little, and they are able to fend for themselves by the age of 6 — 7 weeks. The Nocturnal Curassow is common in the wild, but it is rarely observed because it is active only at night.
Other species of curassows are also found in South America. They have a similar way of life, but they are chiefly diurnal.

HOAZIN
Opisthocomus hoazin

The Hoazin is a most striking and curious bird. It is found in the forests of tropical South America of the Andes, living mainly along the edges of rivers, though it is sometimes seen on wooded mountains. It is 55 — 60 cm long, and both sexes are similar in appearance. Classification of the Hoazin has long been a problem for ornithologists but it has so many unique features that it is now placed in a family of its own. It is a most interesting species with many unusual characteristics. It has a two-part crop, half to store food and half to pre-digest it. The crop occupies a third of the front part of the body. The breast bone and flying muscles are poorly developed, so the Hoazin is not a strong flier. This bird feeds exclusively on vegetable material, particularly the leaves of mangroves, and arum plants.
The untidy nest is made of twigs, and is built on a strong horizontal branch, 2 — 6 m above the ground, and usually overhanging water, where it is protected from predators.
The Hoazin's nesting season covers almost the entire year, the female often incubating two clutches a year. The birds nest in colonies, one male in the company of several females. The hen lays 2 — 4 pale rust-coloured eggs with dark red specks, and sits on them for 28 days. The young hatch naked, but they can see and they move restlessly, staying in the nest for only

a few days. The young birds have a unique feature. On the tip of each wing are two toes with claws, which they use to grasp branches as they creep from the nest. They use the beak and feet when climbing. It is not so far clear whether the toes on the wings are a primitive trait traceable back to the ancient history of evolution, or whether this is a secondary adaptation to life in trees. It is interesting that when they fall into the water, the chicks can easily swim ashore and climb back on to the tree.
After the first three weeks the claws on the wings fall off, the toes become stunted and they no longer exist in the adults. While the young are growing their feathers and are unable to fly the female feeds them on food regurgitated from her crop. After leaving the nest, hoazins stay in groups and their shrill call, something like the crowing of cocks, is often heard.

Hoazin

WHITE-WINGED TRUMPETER
Psophia leucoptera

In the humid forests of the Amazon region, the resonating calls of the male trumpeters are often heard. The White-winged Trumpeter is 46 — 56 cm long and weighs about 1 kg. It is predominantly terrestrial, scurrying nimbly through the thickets. It flies only poorly and for short distances — mainly to reach the branches on which it roosts at night. It is a sociable bird, wandering by day in flocks of a hundred or more. It is often found where there is shallow water, for it is a good swimmer. The loud, strident courting note of the males well resembles the sound of a trumpet. The birds also produce a deep lowing call. A courting male displays the beautiful metallic-green feathers on the front of his neck as he leaps and struts. In the nesting season, usually in March or April, the female builds a simple nest in a hole in a tree near the ground or even on it at the foot of a tree. The nesting depression is sparsely lined, with only a few stalks. The female lays 6 — 10 whitish eggs and incubates them for 23 days. As soon as they are dry the chicks leave the nest. They are already quite independent and their mother merely accompanies and protects them.
Trumpeters eat berries and fallen fruit, as well as worms, molluscs and insects, particularly ants. The Indians keep them as domestic fowls, for they are easily tamed and even make pleasant pets. They also

White-winged Trumpeter

Sun-bittern

serve as watchdogs, loudly announcing the arrival of a stranger.

SUN-BITTERN
Eurypyga helias

The Sun-bittern has an area of distribution stretching from Central America to southern Peru, southern Brazil and northern Bolivia. It has a slender body about 46 cm long, a thin beak and short feet. It lives beside streams in dense tropical forests as well as in mountain situations up to 1 000 m.
In the courtship period, the males display their beautifully coloured plumage in sunlit spots. They stretch their necks, hop, nod and make whistling sounds. These bitterns nest in the rainy season, building large round nests of branches, 2 — 3 m above the ground in trees or bushes. Both partners share in the work, and line the finished nest with leaves and moss, often mixed with earth. The female lays 2 — 3 eggs having yellow-brown ground colour and dark brown and grey dots. Both parents care for their young, feeding them for 3 weeks in the nest, and for a further 10 days when they have left it. By the age of 2 months, the young birds are mature and independent.
Bitterns feed on beetles, caterpillars and molluscs collected from leaves, and they catch crustaceans, fishes, tadpoles and insect larvae in shallow water.

BLUE-WINGED PARROTLET
Forpus passerinus

The Blue-winged Parrotlet is one of the smallest species of parrots, being only about 12 cm long. The female is predominantly blue in colour. This species, widespread in north-eastern Colombia, ranges to Guyana and the Amazon region. It is also found on the islands of Barbados and Jamaica where it has been artificially introduced. It has become common particularly in Jamaica, where it is abundant in southern parts of the island. In Guyana, it is plentiful both along the shore and inland. It exists in 5 species throughout this vast territory.
For most of the year, the birds gather in coveys of about 40. They leave the shelter of the woodland for open country where they peck up grass seeds, and feed on various fruits from trees and bushes. They also raid fruit plantations, for the soft ripe fruit. These birds fly 2 — 3 metres above the ground, in compact flocks.
In the breeding season, which normally takes place in February, but may sometimes be between June and August, parrotlets live in pairs. They usually find a hole deserted by woodpeckers, but if ready-made nesting sites are in short supply, they peck a hole in a termite nest and settle there. The female does not line the nesting depression but lays 6 — 8 white eggs directly in the bottom of the hole. She incubates them for 21 — 23 days, during which time the male feeds her several times a day. Both parents take care of the nestlings for about a month. The fully fledged young then leave the nest and are able to fly, though they are fed for a further 10 — 14 days. Families stay together, joining others to form flocks which roam the countryside.

HYACINTH MACAW
Anodorhynchus hyacinthinus

The Hyacinth Macaw is one of the largest parrots in the world. It is about 1 m long and has an unusually strong beak, with which it easily cracks tough nut-

Blue-winged Parrotlet

shells. Both sexes are the same colour. This parrot is resident in eastern Brazil, in the states of Pará and Mato Grosso. Outside the nesting season, it lives in small family groups of 3 — 6, made up of parents and their offspring. The birds spend most of the time in palm groves, flying from tree to tree in search of food. They mainly eat palm nuts, but they also feed on seeds, and enjoy the fruits of various plants, particularly figs. They unfortunately raid plantations, damaging the crops.

In the nesting season, each pair of macaws defends its own territory. The female usually lays 2 white eggs in the bottom of a hole in a tree, and sits on them for about 27 days, the male feeding her three times a day throughout this period. Both parents look after their offspring, feeding them on soft fruit, and later on seeds. Occasionally they bring them insects. Young macaws leave the nest at the age of 3 months, fully coloured but having somewhat less sheen on their plumage than their parents.

BLUE-AND-YELLOW MACAW
Ara ararauna

The Blue-and-Yellow Macaw covers a range from eastern Panama to Bolivia. It is a parrot of outstanding size, measuring about 85 cm, and both sexes are identically coloured. It frequents the fringes of forests along river banks. For most of the year, the birds stay in flocks, flying from one tree to another looking for seeds, especially those of the Sandbox Tree *(Hura*

Hyacinth Macaw

crepitans). They also feed on palm nuts, and raid plantations, spoiling the fruit crops. They may sometimes feed on large insects.
In the nesting season, which is usually in February, the pairs occupy their nesting territories and look for hollows, particularly in dead palm trunks of the species *Mauritia flexuosa.* The female lays two eggs on the bottom of the nesting cup and incubates them for 24 — 26 days, while the male feeds her on food regurgitated from his crop. The chicks leave the nest at the age of 13 weeks, but are fed by their parents for another 14 days. The families then gather in larger flocks which move throughout the countryside.

Blue-and-Yellow Macaw

SCARLET MACAW
Ara macao

The Scarlet Macaw is another large species of parrot. It is about 85 cm long and both sexes are the same colour. Its area of distribution extends from southern Mexico to Bolivia and northern Brazil. It is abundant and when the birds are not nesting, flocks of more than 200 can be seen in the hills, along the edges of the forests or beside river banks. They form pairs in the breeding season, and often nest close to one another in tree hollows or rocky crevices. Sometimes they make do with holes in high, sandy river banks. The female lays 2 — 4 whitish eggs and sits on them for 25 — 26 days. She leaves the nest for short periods to fly a little, but does not look for food as she is fed regularly by the male. Both parents care for the chicks, and the young macaws, fully fledged and coloured, finally leave the nest at the age of 3 months. They feed on green shoots, seeds and berries, soft fruit and nuts. Occasionally, they catch invertebrates.

RED-BLUE-AND-GREEN MACAW
Ara chloroptera

The Red-Blue-and-Green Macaw is found in forests in a wide belt stretching from Panama southwards across Colombia and Venezuela to northern Argentina and south-eastern Brazil. It is a substantial bird, about 90 cm long, and the male and female are the same colour. Even when they are not nesting, macaws stay near woodland, mainly along river banks. They gather in family flocks or form groups of 20 — 40, and peck seeds, green shoots and various soft fruits among the trees. In the nesting season, they form pairs and defend their territories against intruders. The female usually lays two white eggs and incubates them for about 27 days, while the male feeds her regularly. The young are fed by both parents, and leave the nest when they are 3 months old. Other species of macaw live in South and Central America, though they are not found in such abundance. They all produce a grating, shrill call.

GOLDEN PARAKEET
Aratinga guarouba

The Golden Parakeet is a beautiful bird of north-eastern Brazil. It is about 35 cm long and both sexes are alike, except that the male has a more substantial

Golden Parakeet

Red-Blue-and-Green Macaw

Scarlet Macaw

Green Parakeet

beak. It lives in woodland, particularly near rivers. For most of the year, the birds stay in small flocks, often circling high above the woods. In the nesting season, they form pairs and look for a suitable tree hollow about 15 m above the ground, which they line with slivers of wood. The female normally lays 2 eggs which she incubates for 28 days, while the male feeds her. Both parents look after their young, feeding them for 9 weeks on the nest, and then for a further 3 — 4 days. When they leave the nest the chicks have greenish plumage. They moult into their adult coloration at the age of 15 — 16 months. This parakeet feeds on various seeds, soft fruits and green shoots.

Cuban Parakeet

GREEN PARAKEET
Aratinga holochlora

The Green Parakeet is widespread in Central America. It is found in woodlands on mountains and hills from Mexico southwards to Nicaragua. There are five subspecies, some with red-spotted heads. One subspecies has a red throat and red feathers round the front of its neck. This parakeet is only 32 cm long. For most of the year, parakeets form flocks of 20 — 30, but they are always in pairs. The birds roam throughout the countryside, seeking seeds, fruits and green shoots in the treetops. They also peck at blossoms and eat larvae and insects. In the nesting season, each pair defends the territory around the nest. This is built at least 15 m above the ground in a tree hollow. The female lays only 1 — 2 eggs and incubates them for 25 days, while the male feeds her. The young are cared for by both parents until they are about 8 weeks old, and they leave the nest. If the birds do not find a suitable hole in a tree, they take up residence in a nest abandoned by a bird of prey, settle somewhere among branches, or dig out a hollow in a tree termite nest.

CUBAN PARAKEET
Aratinga euops

The Cuban Parakeet is common in Cuba, but it no longer occurs on the adjacent Isla de Pinos. It is a small bird, reaching a length of 26 cm, and with red-spotted plumage. Some birds have only a few red dots, while others are speckled all over. The Cuban Parakeet is a forest dweller, but when it is not nesting, flocks of these birds visit gardens and city parks to forage for food. They feed on seeds, shoots and fruit, and sometimes raid plantations. The nesting season begins in May, the pairs choosing and defending their territories. They usually find a nest deserted by woodpeckers in the trunk of a palm tree. The female lays 2 — 5 eggs and incubates them for 26 days, while the male feeds her regularly. Both parents look after their young for 8 weeks in the nest and for about 10 days after fledging. Families then merge into flocks.

ORANGE-FRONTED PARAKEET
Aratinga canicularis

The Orange-fronted Parakeet is a small parrot, reaching a length of only 24 cm. It lives in Central America, in a region reaching from western Mexico to western Costa Rica. It frequents woodland, especially in the mountains at a height of 1 000 — 1 500 m. In winter, flocks move to lower situations to find food. They fly from tree to tree seeking seeds, shoots and juicy berries. They particularly enjoy the fruits of fig trees, and often visit orange plantations. Parakeets are noisy and conspicuous birds when they are in flocks, but during the nesting season, when they form pairs, they become very quiet. They move cautiously in the area around their nests, and are difficult to observe. The nest is built in a hollow in a tree, often one abandoned by woodpeckers. The female lays 3 — 5 eggs and incubates them for 28 days, while the male feeds her and guards the nest. The chicks are at first fed on regurgitated food, eating seeds as they grow bigger. They leave the nest when they are 6 weeks old, by which time they have already acquired their adult plumage, though it is less bright than that of their parents.

SUN PARAKEET
Aratinga solstitialis

The Sun Parakeet is distributed throughout south-eastern Venezuela, Guyana and northern and eastern Brazil. It is about 30 cm long and occurs in 3 subspecies, regarded as individual species by some zoologists. The best known of these is the Jandaya Parakeet *(Aratinga solstitialis jandaya).* For most of the year, the Sun Parakeet lives in large flocks of several hundred birds. They frequent thinly forested areas, the edges of primary forests, or places covered with bushes and solitary tall trees in which they can seek shelter in times of danger. They find their food in

Sun Parakeet

Orange-fronted Parakeet

Black-hooded Parakeet

gitated from the crop of their parents, consisting of seeds, soft fruit and insects. The chicks leave the nest at the age of 8 weeks and families gather to form huge flocks.

BLACK-HOODED PARAKEET

Nandayus nenday

The Black-hooded Parakeet is found in south-eastern Bolivia, across the state of Mato Grosso in Brazil, and south to northern Argentina. This parrot, which is about 30 cm long, lives mainly in forested areas beside rivers, and is very abundant in some places. Outside the nesting season, it gathers in flocks of 30 — 40. Together, the birds fly through treetops, looking for various fruits, seeds, and insects. They sometimes settle on the ground among the bushes at the edges of the forests and peck up seeds and berries. In the nuptial period, the pairs seek suitable nesting holes among the trees. The female lays 3 — 4 eggs and sits on them for 25 days while the male feeds her and guards the nest from a nearby spot. The young leave the nest when they are about 7 weeks old, but their parents feed them for another 10 days.

trees and shrubs, occasionally pecking up seeds from the ground. They also feed on shoots and blossoms, and sometimes eat insects and their larvae. Pairs most frequently nest in hollows in palms of the genus *Mauritia.* The nesting period usually begins in February. The female lays 4 — 6 eggs and incubates them for 26 days. The young are fed on food regur-

Maroon-bellied Parakeet

Painted Parakeet

MAROON-BELLIED PARAKEET
Pyrrhura frontalis

The Maroon-bellied Parakeet is a small colourful parrot, which lives in the vast forests and along the rivers of south-eastern Brazil, Uruguay, Paraguay, and northern Argentina. It is only 26 cm long. Outside the nesting season, parakeets live in family flocks or in small groups and roam through the treetops. They are very noisy while in flight, but as soon as they settle and start foraging they become extremely quiet and cautious. They feed mainly on seeds and soft fruit, but occasionally eat insects. The pairs build their nests in hollows in trees, usually those abandoned by woodpeckers. The female lays 3 — 5 eggs and incubates them for 26 — 29 days, the male feeding her three or four times every day. Both parents look after the nestlings. The young birds leave the nest after 43 days, but continue to be fed for another week.

PAINTED PARAKEET
Pyrrhura picta

The Painted Parakeet is widely distributed in the area from Colombia and southern Venezuela in the north to eastern Peru and the northern part of the state of Mato Grosso in Brazil. This parakeet is only 22 cm long. It exists in 7 subspecies, differing in coloration, one having a predominantly red head and yellow ear feathers. The Painted Parakeet is most abundant in coastal woodlands. It feeds on seeds, fruit, insects and larvae, seeking its food in the treetops and rarely descending to the ground. The nest is built in a hole in a tree, over 15 m above the ground. The female lays 4 — 6 eggs and incubates them for 25 days. The young are capable of flight after 6 weeks.

SLENDER-BILLED PARAKEET
Enicognathus leptorhynchus

The Slender-billed Parakeet is about 40 cm long. It has an unusual bill with a particularly long upper part. The male sometimes has a longer bill than the female, but both partners are the same colour. In young birds the upper part of the bill is shorter than in adults. Their plumage is also distinctly darker because their feathers have dark tips. The Slender-billed Parakeet inhabits sparsely wooded mountainous regions of central Chile, and is quite abundant in some areas. It lives among the treetops and often flies to fruit plantations, parks and gardens. After the nesting season, it wanders through the countryside in

Slender-billed Parakeet

Monk Parakeet

large flocks, looking particularly for the seeds of coniferous trees, which it picks adroitly out of the cones. It also feeds on the seeds of wild thistles and raids fields of ripening corn. It also eats fruit and often digs in the soil with its long bill to extract roots, worms, insects and larvae. Farmers dislike this bird and have attempted to eradicate it in many areas.

In November and December this parakeet forms pairs, and nests, usually at an altitude of 2 000 m. The female lays 4 — 6 eggs in a hollow in a tree or in a small cave among the rocks, and incubates them for 27 days. The male feeds her while she is on the nest and after the eggs have hatched he helps her to feed the chicks for 8 weeks while they are in the nest and for 2 weeks after they have left it.

MONK PARAKEET
Myiopsitta monachus

The Monk Parakeet is an inhabitant of eastern Bolivia, Paraguay, and Brazil, occurring particularly in the south of the state of Mato Grosso and in Rio Grande do Sul. Its range also extends south to central Argentina. It is about 30 cm long, and as both sexes are alike in coloration, it is difficult to tell them apart. Four subspecies exist, and blue and yellow varieties which do not occur in the wild have been bred in captivity.

The Monk Parakeet lives in large flocks throughout the year, and during the nesting season it forms colonies of 50 — 70 pairs. The nests are built so closely together on the branches that they combine to form a huge construction sometimes weighing over a tonne, on a single tree. It is by no means unusual for a branch to break and for the nests to be scattered all over the ground. However, the birds rebuild them very quickly, bringing twigs in their bills and weaving them together. In the enormous communal nest, each pair occupies its own compartment, entering by means of an entrance at the bottom which is protected with overhanging twigs so that no bird of prey or wild animal can get inside. As many as 8 such nests may be situated in a large tree, and other colonies build in neighbouring trees. To watch these parakeets building their nests is an unforgettable experience. Some birds carry thin sticks as much as 2.5 m long which swing in the air and are very difficult to control.

The nesting season begins in November. The female lays 5 — 7 eggs and incubates them for 25 — 26 days. Both parents take care of the young, and if the female should be killed, the male is able to rear the offspring. The chicks leave the nest after 42 days, and the parents feed them for a further 10 days. After leaving the nest, the birds may still use it for sleeping. During the nesting period, the male often builds himself a small chamber in which to spend the night. This bird feeds on shoots, fruit and other vegetable material. The Monk Parakeet has been introduced in some parts of the United States and Europe.

ORANGE-CHINNED PARAKEET
Brotogeris jugularis

The Orange-chinned Parakeet has a range which extends from southern Mexico to northern Colombia and northern Venezuela. It inhabits hilly, partially wooded regions, at a height of 500 — 1 400 m, preferring localities with tall isolated trees. This bird is only 18 cm long, and the sexes are identically coloured. Pairs stay together after the nesting season and join other pairs to form small flocks which fly through treetops, making a piercing noise to an-

Orange-chinned Parakeet

Red-fan Parrot

nounce their presence. They feed on nectar which they find in gardens, and also eat tiny seeds and green shoots. In the breeding season, the pairs form colonies, and nest close together in hollows abandoned by woodpeckers, or in the nests of tree termites. The young hatch after 21 — 23 days and are cared for by both parents. They leave the nest when they are about 7 weeks old.

RED-FAN PARROT
Deroptyus accipitrinus

The Red-fan Parrot is about 35 cm long, and has markedly elongated neck feathers which can be erected to form a fan. It is found in the region from south-eastern Colombia and Guyana across to the northern part of the state of Mato Grosso in Brazil. This parrot prefers wooded sites adjacent to savannah grasslands or sandy localities. For most of the year, the birds wander over a large area flying in small groups from tree to tree looking for seeds, various fruit and green shoots. They are also attracted to plantations when the fruit is ripe. At night, they return to areas well supplied with hollows in which to sleep, for each parrot occupies a hollow of its own. In the courtship period, the pairs choose nesting holes over 15 m above the ground. The female lays 2 — 4 eggs and incubates them for about 26 days. The young are fed by both parents, and leave the nest at the age of 2 months. This parrot makes a peculiar wailing cry.

CUBAN PARROT
Amazona leucocephala

The Cuban Parrot is one of the most beautifully coloured species of amazons — stout-bodied green parrots, often kept as cage birds. It is about 32 cm long and occurs in 5 subspecies. Both sexes are identically coloured and are difficult to distinguish. This parrot is confined mainly to Cuba and the Bahamas, but it is also found on smaller adjacent islands. Outside the nesting season, it wanders throughout forests in small groups, and also appears in gardens, parks and fruit plantations where it pecks at the ripe oranges. In the nesting season, pairs take up occupation of their own territories and sometimes fight their rivals. Within their territory they find a suitable tree hollow, usually one deserted by woodpeckers. They do not line it and the female lays 3 — 4 white eggs directly on to the bottom of the hole. She incubates the clutch for 25 days and the male feeds her several

Cuban Parrot

Green-cheeked Parrot

times a day. The newly hatched young are naked and blind, like the young of other parrots. Both parents feed the chicks on a mixture of fruit, seeds and shoots regurgitated from the crop. The young parrots leave the nest at the age of two months when they are fully coloured, but the adult birds continue to feed them for a further 10 days.

Yellow-cheeked Parrot

GREEN-CHEEKED PARROT
Amazona viridigenalis

The Green-cheeked Parrot is a common denizen of north-eastern Mexico, especially of the coastal regions. It is 35 cm long and both sexes are alike in coloration. For most of the year, this parrot is found in large flocks of up to 100 birds. They fly among the crowns of various species of palm trees, pecking the fruit and green shoots. They also feed on soft fruit, seeds and insects. In the nesting season, they form pairs. The female lays 2 — 3 eggs usually in a hole in a tall palm tree and incubates the clutch for 26 days. The male feeds her regularly and both parents look after the chicks. The young parrots leave the nest at the age of 8 — 9 weeks.

YELLOW-CHEEKED PARROT
Amazona autumnalis

The Yellow-cheeked Parrot is one of the most abundant species of amazons. It is widespread from south-eastern Mexico to western Ecuador, north-western Venezuela and north-western Brazil. It is about 35 cm long and ornithologists distinguish 4 subspecies differing in coloration. For most of the year this parrot travels in enormous flocks of up to 600 birds. It lives in areas of woodland, seeking seeds, fruits, shoots and insects in the treetops. The flocks also raid fruit plantations and cause considerable damage to the crops. In the nesting period, the parrots form pairs. The female lays 2 — 4 eggs in a hollow in a tree and incubates the clutch for 25 — 26 days, while the male feeds her. Young birds leave the nest when they are about 2 months old.

YELLOW-HEADED PARROT
Amazona ochrocephala

The Yellow-headed Parrot has a large area of distribution reaching from Mexico to western Colombia, eastern Peru and the north-western states of Mato Grosso and Pará in Brazil. It is also found in Trinidad and occurs in 9 subspecies. It attains a length of 35 cm, and frequents both coastal regions and hills. The female usually lays 4 eggs in an unlined hollow

and sits on them for 29 days. The young leave the nest at the age of 65 days.
Altogether 27 species of amazons are found in tropical Central and South America. Some of them occur at heights up to 3 500 m. Many of them are common, but some species are very rare in the wild, and have populations of less than a hundred birds.

WHITE-BELLIED PARROT
Pionites leucogaster

The White-bellied Parrot inhabits an area from eastern Peru across the region of the Amazon basin to the Atlantic coast. In the south, it reaches northern Bolivia. It is a small species, only about 23 cm long, and both sexes are alike in coloration. For most of the year, the birds roam in small groups and forage for seeds, shoots and fruit in the treetops. This handsome parrot mainly frequents the edges of forests or river banks, and it is quite common in some areas. In the nesting season, it forms pairs and defends its territory against rivals. The female usually lays 4 eggs in a hollow more than 15 m above the ground, and incubates the clutch for 26 — 27 days. The male stays near the nest and feeds the female several times a day. Both parents look after the young, who leave the nest at the age of 10 weeks.

BLACK-HEADED PARROT
Pionites melanocephala

The Black-headed Parrot is found in southern Colombia, eastern Ecuador and north-eastern Peru. In the east it reaches as far as Guyana and the Brazilian state of Pará. Both sexes are about 23 cm long and are identically coloured. This species is abundant particularly in forests in sandy localities and in savannah woodland. It lives in the treetops, roaming in small groups made up of a few pairs, and feeding on seeds, fruit and shoots. In the breeding season, the birds build nests in holes in trees. The female lays 2 — 3 eggs and sits on them for 26 — 27 days. The young parrots leave the nest after 71 — 73 days and their parents feed them for a further 10 days.

Yellow-headed Parrot

White-bellied Parrot

Black-headed Parrot

Spectacled Owl

Bare-legged Owl

SPECTACLED OWL
Pulsatrix perspicillata

The Spectacled Owl is distributed throughout the eastern part of the South American continent, from southern Mexico to Argentina and Bolivia. It reaches a length of 40 — 48 cm, and the two sexes are alike in colour. The Spectacled Owl prefers damp lowland woods and river banks, where it catches crabs, seizing them when they crawl out of the water. It can also be found in mountain areas, up to a height of 1 700 m. In forested areas this owl hunts large insects, small mammals, and lizards — particularly young iguanas.
In the courtship period, these owls make a peculiar sound, similar to the drumming noise produced by woodpeckers. The female lays two white eggs in a hollow in a tree, usually over 20 m above the ground. She sits on the clutch for 28 days and the male brings food to her. For the first ten days after the chicks have hatched the male brings food for the whole family. As the chicks grow bigger the female also begins to hunt. During the day, she often feeds the young on food saved from the previous night's catch. After about 35 days, the young owls leave the nest and perch on nearby branches where the parents bring food to them. After a further 14 days, they begin to fly and learn to hunt.

BARE-LEGGED OWL
Otus lawrencii

The Bare-legged Owl is a small owl, widespread in Cuba, and also on the adjacent Isla de Pinos. It is about 22 cm long and both sexes are identically coloured. It inhabits sparse coniferous and deciduous woods, sheltering by day in holes in trees, often in nests abandoned by woodpeckers. Only after sunset does it set out to hunt, flying noiselessly among the branches and seeking small birds which it skilfully grasps in its long claws. It can catch a bat in flight, and insects on the wing, particularly flying beetles. Like other owls, it regurgitates pellets of undigested food, such as the wing-cases of insects, or bones and feathers of vertebrates. These pellets accumulate on the ground so the nest can easily be found. In the nesting season, the Bare-legged Owl forms pairs. The female lays 2 white eggs in a tree or rocky hollow, and incubates them for 28 days. The male brings food which he gives to her near the nest. Both parents take care of their young, who leave the nest when they are one month old.

Black-banded Owl

BLACK-BANDED OWL
Ciccaba huhula

The Black-banded Owl lives in the dense forests of the Amazon region, mainly along the rivers. It is about 35 cm long. It is solitary for most of the year, hiding by day in the dense foliage of the canopy. It sets out to hunt after sunset, flying almost noiselessly and seeking small sleeping birds. It also hunts small mammals and tree reptiles, especially iguanas. Sometimes it catches large insects, particularly beetles.
It forms pairs in the breeding season, either building a nest in a tree or settling in a nest deserted by some other bird of prey. The female usually lays two white eggs and incubates them for 27 days. During this time the male stays near the nest and brings food for his mate at night. In the first days after hatching the mother never leaves her young, so the male has to find food for the entire family. The young owls are actually fed only by the female, who divides up the prey and gives it to them. Young owls leave the nest when they are 4 weeks old, and after another 2 weeks they are able to fly.

Striped Owl

STRIPED OWL
Rhinoptynx clamator

The Striped Owl is found in the vast forest areas from eastern Mexico southwards to Brazil and Uruguay, being very common in some localities. It does not occur in the Amazon region or in the Andes. This owl is about 35 cm long. It is usually found in thin woodland or along the edges of the primary forests, and it visits large citrus plantations where it feeds on the small rodents which are found there in abundance. After sunset this owl flies noiselessly, low down among the tree trunks, in search of prey. During the day these owls rest in the trees, and when they are not nesting, they often live in groups of more than 15. In the nesting season they form pairs, and the female builds a nest on the ground among thick clumps of grass. She lays about 5 white eggs

Oilbird or Guacharo

and sits on them for 27 days. Although this beautiful owl is relatively common, no details of its nesting habits are known. In the courtship period, it makes piercing, shrill sounds.

OILBIRD or GUACHARO
Steatornis caripensis

The Oilbird or Guacharo is one of the most remarkable representatives of the realm of birds. It is a relative of the nightjars, reaches a length of 43 cm and is similar to owls in its coloration, except for its red, shiny eyes. It is found in mountain forests in Guyana, Venezuela, Colombia, Ecuador and Peru, and along the rocky coasts of Trinidad. It is nocturnal, living by day in large communities deep inside caves in the mountains or along the coasts. It hides in the darkest corners where light never penetrates and is able to fly in pitch darkness among rocky projections and along the twisting passages without hitting them. These birds are endowed with the same capacity as bats. They find their way by listening to the echo of sounds which they emit as they fly and do not need to be able to see. At twilight, flocks of oilbirds leave the caves to forage for food. They live almost exclusively on the oily fruits of palm trees, tearing them off with their hooked bills, and travelling as far as 100 km to find ripe fruit. Their legs are adapted for hanging and climbing in the crowns of the oil palms and on rocky cave walls, and they wave their half-spread wings to help them keep their balance. On very dark nights, they look for food using their sonar devices, but they seek the palm fruits mainly with their eyes. At dawn, the flocks return to their caves, where layers of undigested seeds accumulate in the droppings beneath their roosting sites.
The oilbirds nest in the caves, building nests in crevices in the rocks often hundreds of metres from the entrance. The nest is a flat construction made of crushed fruits stuck together with gluey saliva, and it may be used for several years in a row. The nesting period can be at any time of the year. The female lays 2 — 4 white eggs, pointed in shape so that they are less likely to fall out of the nest, and both partners share in incubation for 33 days. During this time, the eggshells become stained with oil from the birds' food. The chicks are also fed on the rich oily fruit by both parents, and do not leave the nest until they are 120 days old. By this time they have become enormously fat and are nearly twice as heavy as the adults.
The Oilbird was discovered by Alexander Humboldt in 1799 in caves in Venezuela, but it had long been known to the local Indians. They used to go to the dark caves — and sometimes still do — to collect the fat young birds which they used as a source of cooking oil.

Tufted Coquette

TUFTED COQUETTE
Lophornis ornata

The Tufted Coquette is a beautiful hummingbird, 7 cm long, which lives in the woodlands of Venezuela, Guyana and Trinidad. Unlike the male, the fe-

male has no colourful, long tail feathers. In the rainy season, when the nuptial displays take place, two males often confront one another in the air, attacking with their beaks and make piercing cries. They are aggressive even outside the nesting period, chasing rivals from the territories where they find the particular blossoms from which they suck nectar with their tubular beaks. When pursued, these small birds can reach a speed of 80 km per hour, beating their wings 75 times in a second, and during the nuptial flights, the beat increases to 200 per second. This creates the humming sound which gives these birds their name. Hummingbirds are extremely agile and react with quite extraordinary speed. They can also hover motionless in the air before moving off in any direction, including backwards. During courtship, the males try to attract the females by fast flying, and by displaying their colourful plumage. They sometimes bring presents of tiny insects, but when mating is over, each male leaves his mate and looks for a new one while the female begins to build a small nest. She weaves it in the shape of a tiny basket, using cotton-like plant fibres and spiders' webs. Into its walls, she inserts pieces of bark and lichen from the tree in which the nest is situated to conceal it from predators. She lays 2 relatively large white eggs and incubates them for about 15 days. She has frequently to leave the nest to find food, regurgitating nectar and insects from her crop for the chicks. Nectar supplies the enormous amount of energy needed by the adult hummingbirds, and insects found in the flowers provide them with protein. If man lived at the same fast rate as a hummingbird, he would have to consume 130 kg of food every day! When feeding two nestlings, the female actually requires three times the effort! She feeds them for 3 weeks in the nest before the young fly off to fend for themselves.
Hummingbirds are preyed upon chiefly by owls, which attack them at night, and by snakes and large spiders.

RED-TAILED COMET
Lesbia sparganura

The Red-tailed Comet is widespread in Bolivia, northern Argentina and Chile. The male is characterized by a long, fork-shaped tail, which together with the body measures 25 cm. The tail reaches its full length when the bird is 5 — 6 years old. The female is only about 16 cm long. The colourful plumage of this hummingbird has a metallic sheen which changes in tone as the light is refracted in the feathers. A male in flight resembles a huge fast-moving dragonfly. The comet is expert at catching tiny flying insects, and also collects them from the blossoms where it sucks nectar. Nectar is drawn in through the long tongue, which is divided into two thin tubes and works like a lemonade straw. Sucking hummingbirds never settle on a flower to feed, but whirl their wings in a figure of eight pattern as they hover. They can move away backwards from a blossom, which is something no other bird can do. In the courtship display, the male comet extends his tail feathers and turns around in front of the female to attract her attention. After mating he leaves her, and the female builds a fine, deep nest on a horizontal branch. She lays 2 eggs, incubates them for about 16 days, and rears the young, which begin to fly when they are 3 weeks old.

Red-tailed Comet

Crimson Topaze

CRIMSON TOPAZE
Topaza pella

The Crimson Topaze is found in the hot, humid forests of Guyana, Surinam, eastern Ecuador and adjacent parts of the Amazon region. The male is 22.5 cm long and has two markedly elongated tail feathers. The female is only 15 cm long, predominantly bronze coloured, but greenish above and reddish below. This species is one of the most strikingly coloured of the hummingbirds. It lives in primary forests and seeks food in the tallest treetops, often settling high in the branches where it skilfully catches flying insects, and from where it makes short flights to find flowers rich in nectar.

In the courtship period, the males are highly aggressive and fight each other. They not only chase away their rivals, but attack other species of birds often more powerful than themselves. The nesting habits of the Crimson Topaze are similar to those of related species.

Many species of hummingbirds live in captivity, although it is very difficult to catch and transport them without hurting them. Bird-catchers have found a way of trapping even the beautiful Crimson Topaze. They push two tall sticks in the ground in a suitable spot in a meadow. Supported on the sticks is a glass vessel containing artificial nectar made of sugar and honey. The daring birds, relying on their speed to get away, allow themselves to be lured by the bait and begin to suck the sweet juice. The catchers reach out and touch them with long sticky rods to which the birds get stuck. They are then transferred to tiny cloth bags in which they are unable to move and so cannot exhaust themselves. The bags containing the birds are packed into boxes for transportation to other parts of the world. The birds have to be regularly fed on sweet juice throughout the journey.

Streamer-tailed Hummingbird

STREAMER-TAILED HUMMINGBIRD
Trochilus polytmus

The Streamer-tailed Hummingbird inhabits both humid mountain forests and cultivated regions in Jamaica, being found mainly in fruit plantations. The male measures up to 25 cm in length, and has two long tail feathers. The female measures only 10 cm and is bronze-green above and white below, and lacks the long tail feathers. The colour of the beak in males varies with locality, hummingbirds from western parts of Jamaica having red beaks, while those from eastern areas have black ones. This hummingbird makes a loud call when in flight, and like other species of hummingbirds, it feeds on nectar and insects caught on the wing.

After the courtship period, the female is left alone to build her nest. It is made from delicate plant fibres, and is attached to vertical branches of trees or bushes. Here it is afforded better protection from predators than on horizontal branches. The female lays 2 eggs, incubates them for 17 days and feeds the young in the nest for 3 weeks.

White-tipped Sicklebill

WHITE-TIPPED SICKLEBILL
Eutoxeres aquila

The White-tipped Sicklebill is distributed from Costa Rica southwards to Ecuador and north-eastern Peru. It inhabits dense tropical forests in the coastal lowlands as well as those up to a height of 1 700 m in the mountains. It measures about 13 cm, including the hooked bill, which is 2.5 cm long. The shape of the bill is well adapted for sucking nectar from flowers of the genus *Heliconia.* In addition to nectar, it also feeds on tiny insects, which it collects on branches and leaves. It has well-developed, strong feet which enable it to rest on the branches when it is hunting insects. Its habits in the breeding period are similar to those of other hummingbirds.

COLLARED, RED-BELLIED or BAR-TAILED TROGON
Trogon collaris

The Collared, Red-bellied or Bar-tailed Trogon lives in mountain forests up to a height of 2 300 m. It ranges from Central America to northern Bolivia and north-western Ecuador, and also lives in Trinidad and Tobago. It measures about 27 cm. The female is less colourful than the male. She has a predominantly brownish back and chest, her abdomen is dull red, and she has a white stripe on her breast. Trogons rest for hours in the dense treetops. They leave them from time to time to hunt insects, spiders, tiny tree lizards or frogs, and they like to peck soft fruit and berries. The male produces wailing cries, and in the nesting season makes various whistling sounds. The nest is built by both partners in a tree, either in a hole, or in a nesting chamber pecked out in an old termitarium. The female lays 2 white or greenish-white eggs, which she incubates alternately with the male for some 17 days. Both parents feed their young on insects and fruit. After about 3 weeks, young trogons leave the nest, but they are fed for a further 10 days.

Collared, Red-bellied or Bar-tailed Trogon

Cuban Trogon

found in abundance in some places, living in dense treetops. Although it is strikingly coloured, it blends well with the green background. It often perches motionless on a branch, but as soon as it sees a flying insect, it comes to life and sets off in pursuit of its prey. It also catches spiders and tree lizards, and eats sweet fruit. It the nesting season, the males make loud, drawn-out cries to announce their presence. In April, pairs build a nesting chamber in the hollow of a deciduous tree or in an old termite mound. The female usually lays 2 off-white eggs directly on to the bottom of the hole. Both partners share in incubation for 17 — 19 days and take care of the young, feeding them on insects, larvae, and later, on fruit. At 15 — 17 days of age, trogons are fully fledged and grown. They leave the nesting hollow but are fed for a further 10 days, since they cannot catch insects right away. Only then do they become completely independent.

CUBAN TROGON
Priotelus temnurus

The Cuban Trogon is a colourful bird 30 cm long. It is found in Cuba, where it is strictly protected. It inhabits both lowland and mountain forests, and is

CUBAN TODY
Todus multicolor

The Cuban Tody is a small relation of the kingfishers. It is confined to the islands of Cuba and Isla de Pinos, where it frequents sparse woodland or bush-covered banks along the rivers and coasts. It is about 11 cm long and both sexes are alike in coloration. The Cuban Tody is not a timid bird. It is not afraid of man, and even approaches intruders to observe them. It is extremely agile and constantly on the move. Although it does not fly very quickly, it can catch flying insects. While waiting for its prey it sits pressed to a branch, but as soon as a butterfly appears, it takes off, grasps the insect and swallows it. It consumes a large quantity of insects every day, and sometimes catches small lizards among the trees. Outside the nesting season, it is usually solitary. In May or June, todies form pairs and build nests in the clay or sand of river banks. Sometimes, they merely settle in a hole in the ground but in banks they dig out corridors up to 2.5 m long, and with many curves. Even when the construction is only 50 cm deep, it cannot be penetrated by enemies. The female lays 2 — 6 shiny, white eggs. Both partners share in incubation for 20 days and feed the chicks on insects. At the age of 4 weeks, the young leave the nest and learn to hunt independently.

Cuban Tody

Four other species of todies live in Jamaica, Puerto Rico and Haiti, having the same habits as the Cuban Tody.

BLUE-CROWNED MOTMOT
Momotus momota

The Blue-crowned Motmot is a colourful bird, 38 — 41 cm long, which is related to the kingfishers. It is distributed in Central America and the northern part of South America, and also occurs in Trinidad and Tobago. It is mostly found in the rainforests, but it also frequents areas covered with tall shrubs, in both lowlands and mountains. Motmots usually live in pairs throughout the year, the partners often perching motionless together on branches, waiting for prey. After every foray in search of food, they return immediately to their favourite perch. They feed on large insects and tiny lizards and on fruit.
In the courtship period, motmots utter hollow sounding notes. They make careful nesting preparations and can take as long as 10 weeks to dig a corridor over 1 m long in a sand or clay bank. It has several curves and is terminated by a nesting chamber where the female lays 3 shiny white eggs. She and the male take turns in incubating them for 21 days. The young are fed on insects and later on soft fruit. At the age of one month, the fledglings fly from the nest.

Blue-crowned Motmot

Toucan Barbet

TOUCAN BARBET
Semnornis ramphastinus

The Toucan Barbet lives in the mountain forests of Colombia and Ecuador. It is related to the woodpeckers and reaches a length of 20 cm. It is characterized by a powerful beak, which it uses to drill holes in dead trees. As its beak is less strong than that of the woodpecker, it has to look for weak spots in which to drill. It never lines the nesting cup, and the female lays 2 — 5 white eggs directly on fragments of wood in the bottom of the hole. Both partners incubate the eggs for 2 weeks. The young are hatched blind and naked, not gaining their sight for over a week. They are totally dependent on the care of their parents, who feed them on tiny insects, and later on fruit. The parents throw the droppings out

Chestnut-capped Puffbird

and always keep the nest spotlessly clean. After a month, the young are capable of flight and leave the hollow, returning at night for a further three weeks before becoming entirely independent. The adult birds feed chiefly on berries, fruit and insects, and occasionally on small geckos or iguanas.

CHESTNUT-CAPPED PUFFBIRD
Bucco macrodactylus

The Chestnut-capped Puffbird is 14 cm long, and is related to barbets and woodpeckers. It ranges from southern Venezuela west to the Colombian Andes and south to eastern Peru and northern Bolivia. It inhabits vast forests where it dwells in the treetops. It is solitary for most of the year, perching immobile on dry or thinly leaved branches waiting for insects to pass, and darting rapidly after them. It preys particularly on butterflies, beetles and wasps, and after seizing them, it returns to its perch, puffs out its feathers and becomes inert again.
These birds form pairs in the breeding season, and dig out a nest in the sand of a river bank. It consists of a burrow, up to 2 m long, terminated by a nesting chamber. The female lays 2 — 3 white eggs and incubates them with help from the male for some 16 days. The parents look after their chicks for 3 weeks in the nest and then feed them for a further 14 days.

ACORN WOODPECKER
Melanerpes formicivorus

The Acorn Woodpecker is well known for its habit of storing food. It is distributed in a zone ranging from the south-western coast of the United States across Mexico and Central America to the Andes and Colombia. It is about 25 cm long, and the male is distinguished by a black stripe across the head. In northern regions, the Acorn Woodpecker lives in oak forests or in mixed forests containing solitary oak trees. For most of the year, this bird travels in small flocks of 5 — 6, foraging for acorns, nuts, fruit, seeds and insects. In autumn these woodpeckers drill small holes in the bark of trees, especially pines, and stuff an acorn in each hole. If they cannot find any acorns, they store nuts and sometimes even tiny pebbles. Some trees look like sieves, for several thousand such holes can be found in a single pine. Sometimes woodpeckers make their winter stores in log cottages in mountains or in wooden sheds. These rich stores are not used by the woodpeckers alone, but are raided by other birds, squirrels and mice. It is interesting that acorn woodpeckers living in southern regions make no such stores.
In the nesting season, pairs look for territories and announce their ownership with loud, parrot-like calls. They drill nesting holes usually in the trunks of pines or oaks, but sometimes in telegraph poles. The female lays 4 — 6 white eggs and both partners take turns in sitting on them for 2 weeks. They feed their young mainly on ants. When they are 4 weeks old, the young woodpeckers leave the nest, but return at night. After the nesting season, the families stay together as they travel through the woods.
Some parts of the tropical forests of South America are inhabited by the White Woodpecker *(Leuconerpes candidus).* Its habits are similar to those of other woodpeckers.

IVORY-BILLED WOODPECKER
Campephilus principalis

The Ivory-billed Woodpecker is the largest species of woodpecker in the world, but unfortunately it is also one of the rarest. It is about 50 cm long, the male having a red crown, and the female a black one. It has recently been observed several times in Cuba,

Ivory-billed Woodpecker

Acorn Woodpecker

White Woodpecker

and five specimens were reported in Texas in 1961. This bird was once plentiful in the woodlands of Cuba and north-western Mexico, ranging to south-eastern Texas and Florida, but it was hunted for meat almost to the point of extinction. It was always easy to find because its loud powerful call could be heard as far as 1 km away. During most of the year, the Ivory-billed Woodpecker wanders over a wide area within its habitat. In spring it occupies and defends an extensive territory. Here the partners drill a hole 1 m deep in a tree trunk, the female being the more active. This is a demanding undertaking, and can take as long as 14 days, with splinters 15 — 18 cm long flying around. No wonder then that older pairs often use the same nest for several years. The female lays 3 — 4 white eggs at the bottom of the nest, and incubates them with help from her mate, who relieves her mainly at night. The young are fed on insects and larvae. The woodpecker also feeds on a variety of nuts, fruits and seeds. The fledglings leave the nest at the age of 4 weeks and become independent within a further 14 days.

TOCO TOUCAN
Ramphastos toco

The Toco Toucan is the largest species of toucan, reaching a length of about 65 cm and having a huge beak. The beak is very light for its size because the horny outer case is filled with tiny air chambers in a network of fibres. The Toco Toucan inhabits the vast South American forests ranging from Guyana to northern Argentina. For most of the year, toucans stay in small groups of 4 — 5, and their hoarse call can be heard from the treetops. They seldom visit the ground, except to peck at fallen berries or fruit, preferring to find fruit in the trees. Their plant diet is occassionally supplemented with tiny birds, frogs and young iguanas. Toucans also catch insects and invertebrates, and take birds' eggs. At night, they find hollows to shelter them from enemies, especially owls, or hide in dense foliage. When roosting, they put their beaks on to their backs and fold their tails backwards towards the head, ruffling their plumage and looking like feathered balls. During the day,

Toco Toucan

toucans often fall prey to raptors, although they try to escape them in the treetops.
Toucans nest in holes in trees. The female, which is the same colour as the male, lays 2 white eggs, and both partners take part in incubation for 16 days. The young birds have only short beaks which take several weeks to reach their full length. They are fed on insects and soft fruit for about 7 weeks before they leave the nest and fly with their parents. Indians hunt these birds with blowpipes, for they are good to eat.

SULPHUR-BREASTED TOUCAN
Ramphastos sulfuratus

The Sulphur-breasted Toucan lives in rainforests ranging from Mexico to northern Colombia and Venezuela. It occurs both in lowlands and in mountains up to a height of 900 m. It measures 46 — 51 cm, including the beak which is 14 cm long, and both sexes are identically coloured. Outside the nesting season, toucans wander through the treetops in small flocks and forage for food. They feed chiefly on berries and fruit, but also hunt for large insects, especially cicadas, and occasionally catch tree lizards or take nestlings. During the day, they look out for large tree hollows filled with water where they enjoy having a bath.
In the nesting season toucans form pairs, feeding each other and making repeated grating call notes. The nest is made in a tree cavity, sometimes lined with green leaves. The female lays 2 — 4 white eggs, which are incubated by both partners for 16 days. The young are fed by both parents and leave the nest at the age of 6 — 7 weeks.

PLATE-BILLED MOUNTAIN TOUCAN
Andigena laminirostris

The Plate-billed Mountain Toucan is widespread in mountain forests of the subtropical zone of Colombia and Ecuador. It is about 45 cm long and both sexes are alike in coloration. For most of the year, toucans stay in flocks roaming throughout the countryside and pecking the fruits of trees growing beside mountain streams. In the nesting season, they form pairs. Their way of life does not largely differ from that of other species of toucans.

CHESTNUT-EARED ARAÇARI
Pteroglossus castanotis

The Chestnut-eared Araçari is found in forests in the western part of the South American continent, both

Plate-billed Mountain Toucan

Sulphur-breasted Toucan

Chestnut-eared Araçari

in the lowlands and in mountains up to a height of 1 300 m, in an area extending from Colombia to Peru and northern Argentina. It is 35 cm long and both sexes are alike in coloration. For most of the year, araçaris live in small groups among the trees, collecting fruit and insects, which they feed to their young. They sometimes catch small birds and young geckos and iguanas. Before dark, these birds gather at permanent sites, often hollows deserted by woodpeckers, as many as 6 araçaris sleeping in a large hollow. Before sleep, they fold their tails over their backs and push their beaks into their back feathers.

At the onset of the nesting season, the flocks divide into pairs each of which finds and occupies a territory of its own. The nest is built in a tree cavity, usually more than 20 m above the ground. The female lays 2 — 4 white eggs, and both partners incubate the clutch for 17 days and care for the young birds. After about 7 weeks, the fledglings leave the nest but return at night for another 3 weeks.

Crimson-rumped Toucanet

CRIMSON-RUMPED TOUCANET
Aulacorhynchus haematopygus

The Crimson-rumped Toucanet is 28 cm long, and is a relative of the toucans. It is found throughout the Amazon region, in the lowlands, and in mountain localities up to a height of 2 000 m. It forms flocks of about 8 birds which forage together in the treetops, feeding on fruit, insects, eggs, nestlings, and small reptiles. In the nesting period, it forms pairs and builds a nest in a hollow abandoned by woodpeckers, some 30 m above the ground. Toucanets sometimes bore out a hollow for themselves, but only in dead and rotting trunks. The female lays 3 — 4 white eggs and the partners incubate them alternately for 16 days. The young hatch naked and blind, opening their eyes after 16 days by which time they have grown their feathers. They are fed on insects and fruit by both parents. At the age of 43 — 46 days the young leave the nest, but return to it at night for several more weeks.

RUFOUS-TAILED JACAMAR
Galbula ruficauda

The Rufous-tailed Jacamar is a distant relative of the woodpeckers. It is distributed from Central America through South America, south as far as north-eastern Argentina, and west to north-western Ecuador. It also occurs in Trinidad and Tobago. Its total length is 23 — 28 cm, including the beak, which is approximately 5 cm long and the tail, which measures 13 cm. The female has a yellowish throat and a dull yellow abdomen. This species is found in sparse woodland and in the bushes along the edges of primary forests both in the lowlands and up to heights of 1 000 m. Outside the breeding season it is often solitary, though this is not always the case, some pairs not breaking up when the nesting season is over. Jacamars choose a permanent site on a branch on which to perch and from which they suddenly take off to pursue their prey. They are fast, skilful fliers, seizing even dragonflies with ease. They prey mostly on large butterflies, such as those of the genus *Morpho,* and having caught their prey, the birds return to their perch and tear off the butterfly's wings before eating it.
In the courtship season, jacamars make weak, indistinct trilling sounds. During the nesting period, which takes place from March to July, each pair builds a tunnel, 2 — 3 m long, terminated with a nesting chamber, in a sand or clay bank. The same nest is often used for several years. The female lays 2 — 4 shiny white eggs and sits on them alternately with her mate, although she is more patient and at night she incubates the clutch by herself. The male sometimes feeds her during the day. The young hatch after 19 — 23 days and are at first blind and featherless. Both parents feed them on insects for 20 — 26 days on the nest. After leaving the nest, the young stay with their parents for several weeks.

PLAIN XENOPS
Xenops minutus

The Plain Xenops is a tiny bird, only 13 cm long, which occurs in a territory ranging from Mexico to north-eastern Argentina. It dwells in lowland and mountain forests up to a height of 1 700 m. It is also found in plantations, city parks and gardens. It is

Rufous-tailed Jacamar

solitary for most of the year, wandering in the treetops and feeding on insects and larvae. It is particularly fond of ants, pursuing them and feasting on them as they migrate in huge numbers. It also eats small berries. At night, it sleeps in tree cavities.
In the nesting season, usually from December to May, the Plain Xenops forms pairs. The birds either find a hollow deserted by woodpeckers, 2 — 10 m above the ground, or make do with a hole in a gnarled branch. The nest is lined with pieces of bark, and may be used for several years. The female lays 2 shiny white eggs and sits on them alternately with her mate for 15 — 17 days. The young are fed by both parents for 13 — 14 days in the nest, and for a further 10 days after leaving it. The pair often rears another brood in the same nesting season.

BLACK-FACED ANTTHRUSH
Formicarius analis

The Black-faced Antthrush has a very short tail and long legs. This is characteristic of birds which prefer the ground to the trees. It is a fast runner, taking long steps and moving easily through the thick undergrowth. It lives in the tropics of Central and Southern America and in Trinidad, being found in woodland areas from the lowlands to mountain elevations of up to 1 700 m. It is about 19 cm long. It feeds predominantly on insects found under fallen leaves, but it also catches other invertebrates and tiny young reptiles. It pursues migrating ants which move in millions through the countryside, not to eat the ants themselves, but to consume the victims of their migrations. The enormous mass of ants drives out a variety of insects and small animals in its path. The ants devour some, but kill or wound others, and it is these which fall easy prey to antthrushes.
The nesting season takes place in March, the pairs building a flat nest made of thin roots and leaves, up to 4 m above the ground in bushes or trees. The female lays 2 white eggs, which are incubated by both partners, though largely by the female. The young hatch after 20 days and are fed by both parents. They leave the nest at the age of 18 days.

Plain Xenops

Black-faced Antthrush

Barred Antshrike

BARRED ANTSHRIKE
Thamnophilus doliatus

The Barred Antshrike is distributed from Mexico to Guyana and Colombia. This bird, which is about 16 cm long, prefers sparse woodland or scrubland, usually in humid localities. It hunts insects and their larvae in the bushes and on the ground, and seeks anthills from which it takes the inhabitants. It moves nimbly on the ground and moves easily through the undergrowth. In the courtship period, the male produces a loud trilling call from his shelter in a thicket, to attract a mate. The pair builds a deep nest made of grass and plant fibres on a branch. The female incubates her clutch of 3 — 4 speckled eggs for 15 days, and the male occasionally relieves her. The young are fed on small insects and larvae by both parents. The female differs from the male in coloration, being predominantly russet.

LONG-TAILED MANAKIN
Chiroxiphia linearis

The Long-tailed Manakin inhabits forests and mangrove swamps in Central America, from southern Mexico to Costa Rica. The male's tail feathers are 10 cm long, his total length reaching 20 cm. The female is unimpressive in appearance, being green with lighter underparts. The habitats of the Long-tailed Manakin range from coastal lowlands to mountain situations up to 1 500 m. It dwells in treetops or tall bushes and feeds on tiny fruits and insects collected as it flies.

The courtship period is from April to May, at the beginning of the rainy season. Two males undertake their nuptial displays in front of one female. Under the trees, on a spot only about 25 × 50 cm, the two rivals jump up, run around each other, hop on low-situated branches and produce miaowing sounds. These displays take several weeks, and the males have no time for parental duties. The bowl-shaped nest is made of leaves, moss and ferns, and is hung on a branch 2 — 3 m high. It is built entirely by the female, and she incubates her 2 white eggs for 18 — 19 days with no help from her mate. The young hatch naked and blind, and their mother feeds them on a regurgitated mixture mostly of insects, from her crop. After a few days, when the young birds have gained their sight and grown feathers, the female brings whole insects for them to eat. Young manakins leave the nest after two weeks.

WHITE-BEARDED
or BLACK-AND-WHITE MANAKIN
Manacus manacus

The White-bearded or Black-and-White Manakin lives in dense forest undergrowth in tropical regions of South America from Colombia to north-eastern Argentina. It measures about 11 cm. The female is olive-green above and yellow-green below. This species nests at all times of the year. A courting male selects a display territory rather more than a metre square, and containing 2 — 3 small saplings. He flies from sapling to sapling and produces buzzing sounds by shaking his wing feathers. He jumps up on to the branches for the second stage of the nuptial display, and the female also takes part in the courting dance. After mating, the male abandons his partner and finds a new mate. The female alone builds a deep nest on the top of a bush, up to 2 m above the ground, using moss, twigs and leaves. The nest is usually situated on a branch overhanging water. The female lays 2 white eggs, incubates them for 18 —

19 days, and feeds her young for 14 days in the nest and for another 10 days until they are able to fend for themselves. These birds feed on small fruits plucked in flight, and on flying insects.

SPANGLED COTINGA
Cotinga cayana

The Spangled Cotinga lives in South America, in a region from Guyana to northern Bolivia and reaching to eastern Colombia and Peru. It is 21 cm long. The female is blackish-brown above, and greyish-white to dull yellow below. The Spangled Cotinga frequents the edges of forests, thinly forested areas, woodland savannahs, and mountain localities up to a height of 1 300 m. It is solitary for most of the year. It feeds on soft fruits, insects, and other invertebrates, found in the treetops.

It forms pairs in the nesting season, but the male takes little interest in his family. The female builds her bowl-shaped nest of twigs, moss and leaves high in a treetop. Here she incubates her 2 blue-tinged white eggs for 23 days and looks after her young. The chicks, who are like their mother in appearance, leave the nest at the age of 4 weeks.

BARE-THROATED BELLBIRD
Procnias nudicollis

The Bare-throated Bellbird is a sturdily built bird, about 27 cm long, related to the cotingas. It occurs in forest regions of south-western Brazil, Paraguay and north-eastern Argentina. The female differs considerably from the male in coloration. She is predominantly olive-green above, has a black head, and has underparts which are yellowish with olive-green wavy lines. Young bellbirds of both sexes resemble females. This species frequents the treetops and feeds on berries and other soft fruits. It can swallow incredibly large pieces of food.

In the courtship season, the male's sonorous bell-like voice is often heard. When courting is over, the male allows his mate to prepare the nest and rear the brood. She builds a simple flat nest of twigs, about 7 m above the ground in a tree. Here she lays a single egg, creamy white with dark brown specks, and incubates it for 23 days. She looks after the chick

Long-tailed Manakin

White-bearded or Black-and-White Manakin

Spangled Cotinga

Bare-throated Bellbird

Ornate Umbrellabird

for 33 days, after which the nestling becomes fully fledged and leaves the nest.

ORNATE UMBRELLABIRD
Cephalopterus ornatus

The Ornate Umbrellabird is a curious bird of tropical South America. It is found from Venezuela to northern Bolivia and eastern Peru, where it inhabits mountain forest localities at a height of 1 300 m. It is 40 — 48 cm long. The male's head is adorned with an umbrella-shaped crown 5 cm tall, and his neck bears a 15 — 30 cm long feathered dewlap. The female has duller colours, a smaller crown, and a dewlap only 2 cm long. A courting male stretches his dewlap, pushes it in front of his body and ruffles his plumage. A female watching this impressive courtship is attracted by the irresistible looks of her suitor, and flies to him. The male takes no part in the construction of the nest or in the rearing of the young. The female builds a simple nest of dry twigs in a forked branch, about 10 m above the ground. She sits on her single egg, which is white with dark brown specks, for 22 days. She feeds the chicks for about a month, until the young umbrellabirds leave the nest. These birds feed on berries, soft fruit and insects.

King Flycatcher

KING FLYCATCHER
Onychorhynchus coronatus

The King Flycatcher is found in lowland forests, especially along the edges and in clearings, in an area from Mexico to western Venezuela and northern Colombia. It is about 17 cm long and feeds on insects, usually caught on the wing. It also collects larvae on leaves of trees and bushes. In the nesting season, the pairs very fiercely defend their territories, driving away much more powerful birds than themselves, including raptors. At this time the male tries to attract the female's attention by erecting his colourful crown and ruffling his plumage. The female makes a nest of leaves, moss and fine grass stalks, sometimes in a tree hollow, and sometimes in a deserted burrow, as much as 1 m long, in a sandy bank above the water. She incubates her two eggs by herself for 19 days, which is a relatively long period for such a small bird. The newly hatched young are fed on insects and spiders by both parents. Young flycatchers leave the nest at the age of 22 days, but their parents feed them for another 14 days.

STRIPE-HEADED TANAGER
Spindalis zena

The Stripe-headed Tanager has its home in the Greater Antilles, the Bahamas and the small islands off Yucatan. It reaches a length of 15 — 20 cm. The female has a grey-striped head, olive-green underparts and breast, grey flanks, and brownish-black wings and tail with olive-coloured edges. Young birds resemble their mother in coloration, but the olive-green shade always predominates in young males. The Stripe-headed Tanager is common in sparse woodland, even in the mountains, and it can also be seen on bush-covered slopes, in parks and in gardens. Except in the nesting season, it roams in flocks throughout the surrounding countryside, sometimes visiting Florida. It often raids orange plantations, pecking at the sweet soft parts of ripe fruit. Its diet is complemented with insects, their larvae and spiders. It comes down to the ground to drink and bathe in shallow water or in tanks or pools.

Pairs are formed in the nesting season, the nest being built by the female with only occasional help from the mate. It is made of fine roots, moss, cobwebs, and interwoven pieces of thin bark, and is situated on a branch. The female lays 2 — 3 eggs, off-white in colour, with brown specks and dots, and incubates the clutch for 13 days. The young are fed by both parents for 13 days in the nest and for a further 10 days afterwards.

MASKED CRIMSON TANAGER

Ramphocelus nigrogularis

The Masked Crimson Tanager covers south-eastern Colombia, and spreads southwards to eastern Peru and eastwards to the Amazon basin. It is 20 cm long. The female resembles the male but lacks his pronounced coloration. This tanager frequents woodland with thick undergrowth, usually near streams, both in lowlands and mountains at about 1 300 m. It finds both food and shelter in the treetops. Its diet consists of insects, spiders and juicy fruits. For most of the year, it stays in small groups, but in the nesting season the pairs often form a colony, and build their nests near to one another. The nest is a flat structure with a shallow cup, and is constructed by the female from fine roots, stalks of grass and pieces of bark. She incubates the clutch usually of 2 whitish, brown-spotted eggs, for 12 — 14 days, while the male brings food to her. The young are fed by both parents for 10 — 13 days on the nest and for a further 14 days afterwards until they can find food for themselves.

Stripe-headed Tanager

BAY-HEADED TANAGER

Tangara gyrola

The Bay-headed Tanager, which reaches a length of 14 cm, has a very wide area of distribution. It covers a zone stretching from Panama across tropical South

Masked Crimson Tanager

Bay-headed Tanager

Larger Spotted Tanager

America to eastern Peru and northern Bolivia. It also occurs in Trinidad. It lives in woodlands and among the bushes on the edges of forests, but it also visits cocoa plantations. This species is not confined to the lowlands but can be found at elevations around 2 300 m. Throughout its extensive area of distribution, it exists in several subspecies differing in coloration. The pairs live together all through the year. When the nesting season is over, they wander through the forests catching insects and larvae, and pecking at fruit.

The nesting season varies according to the area of distribution. In Trinidad, it lasts from March to May, while in Colombia it takes place in June. The nest is a flat construction made of roots and moss, lined with fine plant fibres. The female builds it by herself, on a branch 3 — 8 m above the ground. She lays 2 cream-coloured eggs with brown dots, and incubates them for 13 — 15 days. The nestlings are fed by both parents until after 14 — 16 days, the young tanagers leave the nest. The adult birds usually rear a second brood, and some pairs even rear three broods during the year.

Seven-coloured Tanager

LARGER SPOTTED TANAGER
Tangara guttata

The Larger Spotted Tanager is another related species of the rich bird realm of the South American continent. It is distributed in the northern part of South America and in Trinidad. It is about 12.5 cm long. The female resembles the male, but her plumage is predominantly grey, and she lacks the yellow spot on the forehead and across the eyes. She is usually bigger than the male. Such duties as the building of the nest and incubation are left to her. The nest is constructed on a horizontal branch 2 — 8 m above the ground, from stalks, moss and pieces of bark. The male has been seen to help her, but this behaviour is exceptional. The female lays 2 brown-spotted eggs, which she incubates for 13 — 15 days. The young are fed on insects and sweet fruits by both parents. At the age of 14 — 16 days, the young tanagers leave the nest, and the adults rear at least one more brood during the year.

SEVEN-COLOURED TANAGER
Tangara fastuosa

The Seven-coloured Tanager is abundant in the forest regions of the states of Pernambuco and Alagôas in eastern Brazil. It is about 14 cm long, and the female is distinguished from the male by her bluish head feathers and more matt colouring. This bird lives in the crowns of tall trees, where it forages for insects and larvae, especially hairless caterpillars, and pecks at a variety of fruits, mainly soft berries. The nest is built on a branch by the female, and comprises roots, moss and other plant material. She lays 2 brown-dotted eggs and incubates them carefully for 2 weeks. The chicks are fed by both parents, until they leave the nest at the age of 15 days. Families stay together and later merge into large flocks.

RED-LEGGED HONEYCREEPER
Cyanerpes cyaneus

The Red-legged Honeycreeper is found in Central America in the Greater Antilles and almost throughout the northern half of South America except in the western regions. It is a small bird, measuring about

12 cm. The female is dark green above and yellow-green below. Her wings and tail are blackish-brown and her feet are brown. After the nesting period, the male moults into a similar plumage, but his feet are permanently black, which distinguishes him from his mate.
This species frequents sparse woodland and visits parks and gardens. It dwells in the treetops, often in the company of tanagers. It seeks blossoms and sucks out their nectar with its slender bill. Honeycreepers often suck the sweet juice of ripe oranges in plantations, and they also catch insects.
The nest is built by the female in a forked branch of a tree or bush. It is made from grass, coconut and other fibres, and is softly lined. The female incubates 2 — 3 blue-green, russet-spotted eggs for 13 — 14 days, the male meanwhile sitting on a nearby branch or on the edge of the nest. The young hatch out blind, opening their eyes on the ninth day. They are fed chiefly by the female, and leave the nest at the age of 15 — 17 days, although the female feeds them until they are 22 days old. They do not return to the nest but roost on branches, and become totally independent after just over a month.

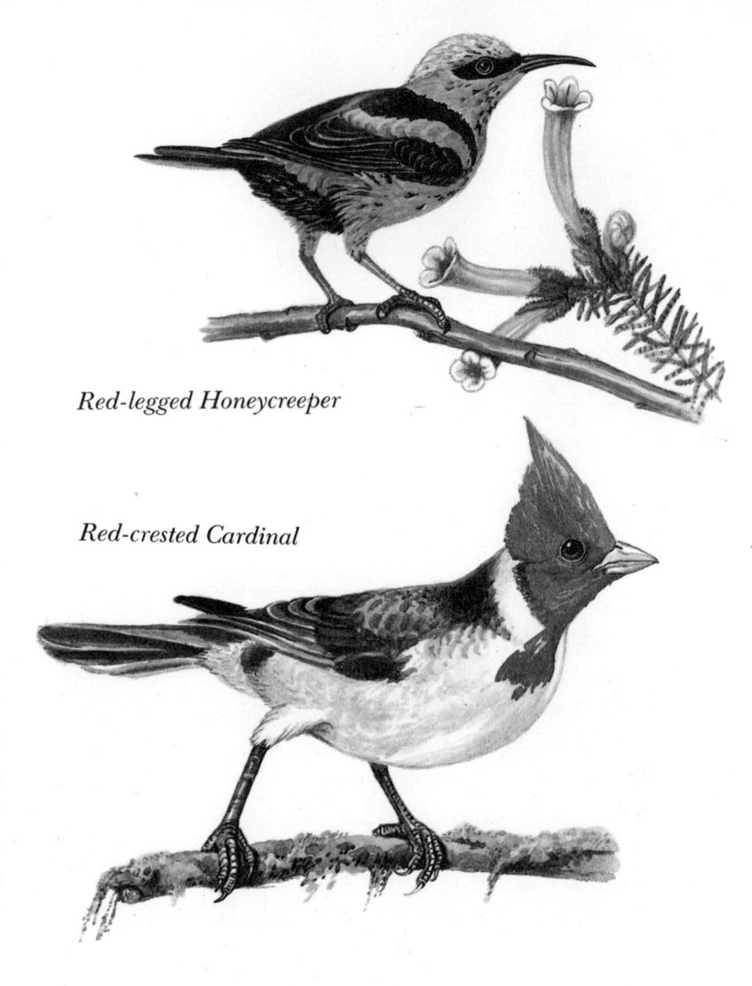

Red-legged Honeycreeper

Red-crested Cardinal

RED-CRESTED CARDINAL
Paroaria coronata

The Red-crested Cardinal is distributed from southern Brazil to Paraguay, Uruguay and Argentina. It is about 19 cm long, and both sexes are identically coloured. It lives on the edges of forests, in damp bush-covered sites with solitary trees, and in parks and gardens. It feeds on seeds, shoots, insects and invertebrates. The nest is built among the dense branches of coniferous trees by both partners, and is made of sticks, stalks, moss and bark. The female lays 3 — 4 white eggs with grey-green and olive-brown spots. She incubates them for 12 — 14 days, while the male feeds her and occasionally relieves her. Both parents look after their young and feed them for 14 — 17 days in the nest and for a further 2 weeks afterwards.

YELLOW CARDINAL
Gubernatrix cristata

The Yellow Cardinal lives in eastern Brazil and eastern Argentina, inhabiting woodland and bushland with occasional tall trees. It often visits plantations

Yellow Cardinal

Rosy Bunting

to collect insects and tiny seeds, especially those of weeds. It also pecks shoots and berries. This bird measures about 20 cm, and the female differs from the male in her grey-white breast plumage.

In the nesting season, each pair defends its territory and drives out intruders. Tiny dry twigs, stalks, grass, moss and plant fibres are used to make the nest which is situated in the bushes. The female works by herself, but the male accompanies her when she is collecting building material, and he also brings her food. She takes about 3 days to build the nest, and afterwards she lays 4 eggs, greenish with glossy black dots, and incubates them for 12 — 14 days. The male stays in the neighbourhood and makes loud calls to indicate that the occupied territory is his. The young remain in the nest for about 12 days, and are looked after for a further 14 days after leaving it.

Montezuma Oropendola

ROSY BUNTING
Passerina rositae

The Rosy Bunting is an inhabitant of wooded slopes in the Mexican states of Oaxaca and Chiapas. It is about 12.5 cm long, and the female has coffee-brown underparts and greyish-brown wings and back. For most of the year the birds live in small groups, but in the breeding period, each pair occupies its own territory. The male energetically chases his rivals away and also repels other species of birds, sometimes bigger than himself. The nest is made by the female among the dense branches of trees or bushes. It is composed of small twigs, grass stalks and moss. The female lays 4 — 5 pale blue, russet-spotted eggs and incubates them for 13 days. She also takes care of the young, the male assisting her only occasionally, for he is responsible for guarding the nest and driving out intruders. The chicks leave the nest when they are about 12 days old, but the parents continue to feed them on tiny seeds, insects and larvae for another 10 days. Buntings also eat fruits, berries, and green shoots.

MONTEZUMA OROPENDOLA
Gymnostinops montezuma

The Montezuma Oropendola is widespread in tropical forests ranging from southern Mexico to eastern Brazil. The male is up to 47 cm long, and the female somewhat smaller. This species has strong claws, which it uses to grip branches while it is hanging on them and sucking nectar or pecking at fruit. It also feeds on insects.

This bird lives in flocks, and nests in colonies among tall trees. While the females are busy building their nests, the males keep guard and warn of the presence of enemies such as birds of prey. Oropendolas often take up residence in a tree which contains a wasps' nest. Predators usually avoid these trees, because the wasps attack intruders very savagely. The female

Blue-naped or Urraca Jay

makes her nest from strips of bark and palm leaves. First, she tears off a piece of a leaf, holds it in her tongue and flies out sideways, tearing away a fibre several metres long. She then weaves numbers of these strips around the end of a branch, and constructs a nest which looks like a hanging pouch, up to 1.5 m long and 15 cm wide. The entrance to the nesting cup is at the top, immediately under the branch. Oropendolas also use fibres from plants of the genus *Tillandsia,* which make the structure firm but flexible. They take 3 — 4 weeks to build, and up to 30 of these conspicuous nests can be seen hanging in one tree. The nesting cup is lined with fine grass. The female incubates her clutch of 2 eggs for 14 days, and rears the brood by herself, because there are relatively few males in a colony. The usual ratio is one male to four females, which is why these birds do not form pairs. The young leave the nest at the age of 3 — 4 weeks, but remain in the colony.

BLUE-NAPED or URRACA JAY
Cyanocorax chrysops

The Blue-naped or Urraca Jay is common in the forests of southern Brazil, Paraguay and Uruguay. It is about 37 cm long and both sexes are identically coloured. The head is crowned with a short, thick cap of feathers which looks rather like a barrister's wig. For most of the year, this jay wanders through the forests foraging for insects and their larvae, spiders, molluscs, and small vertebrates, such as young frogs, iguanas and geckos. Jays also take the eggs and nestlings of small birds, and peck at berries and soft fruits when visiting plantations near the forests.
In the nesting season, the pairs build nests in tall, usually thorny bushes. The nests are made rather haphazardly from dry twigs, grass and moss and lined with fine roots and animal fur. The female lays 6 — 7 bluish eggs and incubates them for 17 — 18 days, occasionally being relieved by the male. The parents feed their offspring on insects, and later on tiny vertebrates. At the age of three weeks, young jays leave the nest, but remain on branches near the nest, where the parents come to feed them for another 2 weeks, even though the chicks have already begun to fly. After becoming independent, young jays stay with the adults and merge into large flocks.

GREEN JAY
Cyanocorax yncas

The Green Jay is distributed in both tropical and subtropical regions, in a zone ranging from south-eastern Texas southwards to northern Bolivia and northern Peru. It lives in forests with thick under-

Green Jay

growth, in lowlands along the coast, and in mountains up to a height of 1 700 m. For most of the year jays live in flocks and are often seen in city parks. They also raid plantations to eat ripe fruit, such as oranges. They also feed on berries, seeds, insects, and occasionally on eggs, nestlings and small lizards.
As soon as the nesting season begins, in April or May, the flock breaks up and the pairs look for nesting territories. The males at this time make various miaowing and other sounds. The nest is built on branches of low trees or bushes, up to 5 m above the ground. The structure, which is made of thorny twigs, lichens and moss, lined with fine roots, is usually well hidden from predators. The female lays 3 — 5 greyish-white eggs, densely covered with brown spots. Both partners share incubation for 17 days, and both feed the chicks. The young remain in the nest for 22 days, and continue to depend on the care of the adults for some time longer. Families later merge to form flocks.

COMMON IGUANA
Iguana iguana

The Common Iguana is the best known and the largest of all the species of iguana. It reaches a length of over 2 m, but most of this consists of a whip-shaped, vertically flattened tail. This iguana inhabits forests throughout tropical Central and South America where it frequents treetops and bushes near water. Its long, slightly hooked claws help it to climb nimbly among the branches. It is even capable of leaping several metres from tree to tree, but only from a higher spot to a lower one. It stays in the trees during the day, but at dusk it searches for food on the ground. Among the greenery of the tropical forests, it blends with its surroundings and is invisible to its enemies. When in danger, it dives underwater and swims using only its tail, and with its legs pressed close to its body. It can remain submerged for several minutes, easily escaping the original predator, but sometimes falling prey to a caiman or crocodile lurking in the water.
The iguana feeds on both animal and plant food. Young iguanas hunt insects, spiders and other small invertebrates while larger specimens take birds' eggs and nestlings. Very occasionally iguanas catch fishes. Adult iguanas feed on fruits, leaves and shoots.
In the mating season, the female digs a hole in the sand and lays up to 70 eggs, which she carefully buries. The eggs are about 35 mm long and 23 mm wide and have soft leathery shells. Sometimes several females lay their eggs in the same spot. The length of incubation depends on the surrounding temperature, but young iguanas normally hatch after 65 —70 days. They measure 18 — 23 cm at first, and grow very quickly so that by the time they are 2 months old they already measure about 30 cm, and after two years, they are up to 120 cm long. The females are sexually mature at 3 years of age.
Iguanas are often hunted for their meat. Specially trained dogs are used to track these reptiles in the forest greenery, the hunters following the sound of barking. Iguana eggs are collected for food by local people as well as by a variety of predators.

PLUMED BASILISK
Basiliscus plumifrons

The Plumed Basilisk is one of the most beautifully coloured of the iguanas. It lives in the forests of Costa Rica, and reaches a length of 80 cm, though much of this is its very long tail. The male has a high leathery crest along its back and tail. South America is the home of four other related species.
The Plumed Basilisk frequents bushes and trees beside rivers, and when in danger, it dives beneath the water and swims to a safe place. It finds its food among the branches, catching insects and small vertebrates, and picking sweet fruits.
These sociable animals communicate by rhythmically nodding their heads. Jerking head movements are used by adult males to threaten each other. In the breeding season the female digs a hole in soft soil, using her left and right foot alternately. Here she lays about 13 eggs in soft leathery shells. When the clutch is complete she covers it over, pressing the soil down firmly with her chin. When this exhausting task is accomplished, she leaves the eggs and returns to her tree. The young usually hatch after 80 days. Some have been found to hatch after only 35 days while others have taken as long as 99 days, for the length of incubation depends on the temperature and humidity of the soil. Immediately after birth the young basilisks seek shelter in the bushes. They are dark green, almost black in colour, with pale yellow

Common Iguana

abdomens and light green necks. They begin to assume their adult coloration after 10 days, turning paler, but with brown and black lines and white spots remaining on their flanks. Although this basilisk is absolutely harmless, local Indians fear it, for they believe it is a supernatural creature made up of several other kinds of animals. According to one legend, it obtained its crest from a rooster, its tail from a snake, its predilection for water from a frog, and its eyes from the Devil. Whenever it is seen near a human dwelling, it is believed to foretell the death of a member of the family.

KNIGHT ANOLE
Anolis equestris

The Knight Anole is one of 170 species of small iguanas widely distributed throughout South and Central America. It is the largest of the anoles and is found in Cuba. It measures up to 45 cm, including the extremely long tail. This species lives on the branches of tall shrubs and trees, where it expertly catches insects and spiders, and small vertebrates such as young reptiles or nestlings.
Knight anoles can undergo colour changes in the same way as chameleons, although they are not related to the true chameleons of Africa. They all have colourful dewlaps, normally folded under the throat, but extended when they become excited. The colour varies from species to species, that of the Knight Anole being pink to orange. The extended dewlap is visually very striking and is used as a signal by males in the mating season when they are seeking their territory. The males also use it to impress the females and deter rivals. They become very aggressive, and engage in fierce fights, sometimes inflicting serious injuries upon each other. Females have also been

Plumed Basilisk

Knight Anole

seen to fight. The results of these conflicts are revealed by coloration. The winning male remains pale green, while the loser turns brown, showing his rival that he acknowledges defeat.
After the courtship period, the female digs a hole under a bush with her forelegs. Here she lays a single leathery-shelled egg which she covers with soil. She lays a second egg after a period of three weeks, and continues in this way, laying 6 — 10 eggs in the course of the year. The young hatch after 50 — 60 days and immediately climb up into trees and bushes. These anoles often fall prey to raptors and reptiles.

COMMON TEGU
Tupinambis teguixin

The Common Tegu is a heavily built lizard of the tropical regions of South America. It reaches over 1 m in length, almost half of which consists of a long, whip-like tail. It lives mainly in primary forests, but is frequently found in adjacent sugar-cane plantations. It can also be seen in some numbers on the coast, in areas covered in thick brushwood. Each individual digs its own burrow in which to shelter and hide from its enemies. The Common Tegu is preyed upon chiefly by carnivores, including domestic dogs.

Common Tegu

When cornered, it resists bravely, biting ferociously and hitting out with its powerful tail. When running, it sometimes stands up on its hind limbs.
Tegus feed on small vertebrates such as other reptiles and nestlings, and on birds' eggs, insects and molluscs. The adults also eat soft fruit and berries. They are themselves hunted partly because of the damage they cause to poultry and also for their meat. Dogs are used to track and surround them, and they are dug out of their burrows. The fat commands a high price, because it is believed, wrongly, to be a cure for snakebite.
The female lays as many as 50 leathery-shelled eggs about 7.5 cm long in large termite mounds. She digs a hole in the hard wall of the termitarium, catching many irritated termites at the same time. The termite workers soon close the opening, and so protect the eggs from predators and bad weather. After about 50 days the young hatch and crawl out, dispersing in the neighbourhood and digging out their own burrows.

FALSE CORAL SNAKE
Anilius scytale

The False Coral Snake has a body oval in section and about 80 cm long, with unelongated abdominal scales. Its skeleton shows traces of reduced hind limbs and a pelvic girdle. This burrowing snake is found in northern parts of South America, in sparse woodland or on the edges of primary forests, and in places overgrown with vegetation. It lives in holes in the ground, beneath the trunks of fallen trees, in decaying leaves, or under the peeling bark of rotting stumps. This snake preys on small lizards, and on other varieties of snakes, pursuing them underground. It strangles its prey in the coils of its body and swallows it whole.
The female gives birth to 5 — 10 live young, which immediately seek their own shelters. At first they feed on worms and insects, but later they catch the young of other reptiles such as anoles.
The red and black rings of this harmless snake resemble the warning colours of venomous coral snakes. It is therefore avoided by predators as well as by the local people.

BOA CONSTRICTOR
Boa constrictor

The Boa Constrictor lives in woodlands and bush-covered regions of South and Central America, being most common in Brazil, Venezuela, Ecuador and Peru. Its ground coloration varies from pink to russet. The largest known specimen measured 5.6 m, but an average boa reaches about 4 m.
The Boa Constrictor is nocturnal, hiding by day in caves, rocky crevices or among tree roots, and setting out to hunt after sunset. In dense primary forests where the light does not reach, it hunts during the day as well. It waits for its prey on trees, being al-

False Coral Snake

Boa Constrictor

most invisible in the shade of leaves and branches. When the prey approaches, the boa swiftly slides from the branch, grasps the animal with its fangs and wraps its coils around it. This happens so quickly that the prey has no time to defend itself. Smaller animals are strangled in seconds, larger ones take a few minutes. When the prey is dead, the boa releases its grip and swallows the entire animal, head first. Boa constrictors usually hunt small vertebrates, such as birds and rats. Large specimens may attack the fawns of small species of deer. They cannot swallow larger animals and are only of danger to men who try to catch them.

The species is viviparous, the female often giving birth to live young in a sheltered place. Sometimes, however, the young may be covered in membranous sacs which the young snakes perforate with their heads. She usually gives birth to 30 or so young — larger females sometimes having more. Young boa constrictors which grow up in captivity soon become tame and even friendly. Boas 3 m or more long, brought up in this way, are harmless even to children. Some Indians keep them in the home like pet dogs, and use them to catch rodents. Venomous snakes are thought to avoid houses where boa constrictors are kept. Some women even place boas next to their sleeping children to protect them from poisonous species. In other regions, these snakes are hunted for their meat and skins, and the fat is used in various medications.

The Boa Constrictor has an average lifespan of 20 — 30 years.

GREEN TREE BOA
Corallus caninus

The Green Tree Boa inhabits the vast woodland areas of Brazil and Guyana. It can reach a length of 1.5 m, but smaller specimens are more common in the wild. This arboreal snake has a stout body and a conspicuously large head, with huge, well-developed jaws and long fangs. It hunts birds killing them with its powerful crushing jaws and then swallowing them whole while holding them in the coils of its body.

This snake occurs chiefly on the edges of primary forests. During the day, it hides in dense treetops, coiled around a branch, and in the evening it begins to forage. In dark, dense forests it also hunts in the late afternoon. Among the green foliage, its body can scarcely be seen, and even its eyes are green. Near villages, it kills many stray chickens and is therefore hunted by local Indians. They are only too well aware of its sharp fangs and its aggressive nature. They never try to catch it by hand, but always use long sticks. Although this snake is not venomous, its bite is painful.

The female bears 15 — 20 live young which immediately climb up on branches and shelter in the foliage. They feed on nestlings.

MUSSURANA
Clelia clelia

The Mussurana is one of the best-known South American snakes. It is distributed from Guatemala to Brazil, and its thick, muscular body reaches a length of 2.5 m. Adult specimens have a glossy blue-black coloration, but the young are a coral-red colour. This snake has venom fangs at the back of the upper jaw, its poison being lethal to small animals. It is a protected species because it hunts the poisonous rattlesnakes which abound in some localities. It is completely resistant to the bite of rattlesnakes, never succumbing even to deep wounds inflicted by their fangs. On seeing a poisonous snake, it immediately attacks, grasping it with its teeth and squeezing it in the coils of its strong body. The prey is paralyzed by the venom, and dies within twenty minutes. The rattlesnake, which may be up to half the size of the killer, is swallowed head first within ten minutes. Small rodents are also caught.
In many parts of South America, attempts have been made to introduce this snake in areas with a high occurrence of venomous snakes, but it has never acclimatized. Although it is strictly protected, it is relatively rare.

SOUTHERN CORAL SNAKE
Micrurus frontalis

The Southern Coral Snake is one of the most colourful but also one of the most venomous snakes. It is found in forested regions of Brazil, Uruguay, Paraguay and northern Argentina and reaches a length of about 1.5 m. The small venom fangs are situated in the front of the upper jaw. They contain a closed groove terminated by a tiny opening and resemble a hypodermic needle. From time to time, the coral snake grows new venom fangs to replace the old ones. The venom is highly toxic, affecting not only small rodents but man as well. The bite of this snake can prove fatal unless an antitoxin is available, though fortunately there have been few cases of men bitten by coral snakes. These reptiles are very timid, hiding by day beneath the bark of tree stumps, under stones or in holes, and setting out to prey after sunset. They glide through crevices and hollows, searching for tiny lizards and snakes. They grasp the body or head of the prey, kill it with their poison and swallow it whole, head first.
The female lays about 10 eggs in leathery shells among the roots of trees or in holes in the ground. The young hatch after 3 months and measure about

Green Tree Boa

Mussurana

Southern Coral Snake

7 cm. Very soon they start looking for their own shelters.
Many other species of beautifully coloured coral snakes live in Central and South America.

BUSHMASTER
Lachesis muta

The Bushmaster is the largest of the rattlesnakes, and can reach a length of up to 3.6 m. This most dangerous snake inhabits the edges of forests and plantations in tropical South and Central America. It has large venom fangs and produces a considerable quantity of strong poison. When disturbed, it rapidly shakes the end of its tail, as is the habit of rattlesnakes, and makes a whirring sound like dry leaves. It is most aggressive, and when approached by man, it rushes forward with its mouth wide-open. However, it never pursues man as it is often believed. If an antidote is not available, the Bushmaster's bite usually proves fatal.
The female lays about 12 eggs under leaves, although related species bear live young. The Bushmaster is nocturnal, hunting after sunset for small rodents.

Bushmaster

SCHLEGEL'S PIT VIPER
Bothrops schlegeli

Schlegel's Pit Viper ranges from Honduras and Guatemala to Colombia, Venezuela and Ecuador. It is a forest-dwelling species, gliding silently among the trees. It grows to a length of 1 m, and although it is a venomous species, its bite is not dangerous to man. It occurs in two colours. Some specimens are green to olive-green with black and red speckles, while others are yellow with sparse black spots. This snake is characterized by several elongated scales like spines above the eyes, for which reason it is sometimes called the Eyelash Viper. During the day, it hides in the thick tangle of green foliage, scarcely visible on account of its protective coloration. After twilight it comes out to hunt, slowly moving along the branches as it searches for tiny birds.
This beautifully coloured snake often reaches Europe in banana clusters, exported in crates from America.
The female produces abour 8 live young which scatter among the branches immediately after birth.

THREE-STRIPED ARROW POISON FROG
Dendrobates trivittatus

The Three-striped Arrow Poison Frog reaches a length of about 4 cm. It is found in primary forests, ranging from Peru to the Amazon region, where it is very abundant. It lives among dense foliage in which

Schlegel's Pit Viper

it easily escapes attention. In the breeding season, it often moves to the ground to display its striking coloration. Despite its conspicuous metallic sheen, it seldom falls prey to other animals but rather repels them. All the frogs of the genus *Dendrobates* secrete a very efficient poison from their skin glands, paralyzing the muscles of other animals. A predatory carnivore or bird, seizing this seemingly defenceless prey, will at the least suffer sharp intestinal spasms after devouring the frog. After an experience of this kind, the attacker learns to avoid this alluring mouthful in future. Smaller animals may die as a result of eating this frog. The venom is most effective when it gets into the blood stream and acts on the heart and nervous system. This can happen as a result of the tiniest scratch. Some people suffer painful inflammation even of healthy skin. South American Indians wrap their hands in leaves when catching these frogs. They hold the frogs over a flame to release the poison from the skin, and use it to coat the tips of arrows used in hunting.

This tree frog has an interesting method of rearing its young. The female lays 6 — 8 eggs in a damp site on land. The male guards the clutch for 10 — 14 days, until the tadpoles hatch out. They immediately crawl on to the male's back and attach themselves to him by means of adhesive discs on their mouths. The male carries the tadpoles wherever he goes, occasionally dipping them into water which has collected in tree hollows. When they are 14 days old, the male submerges in shallow, still water. The tadpoles release themselves and stay in the water for 6 weeks. Here they undergo metamorphosis into adult frogs, feeding on tiny insect larvae, especially those of gnats. When they reach the adult stage, they leave the water and settle in the trees where they catch insects and spiders.

In some species of the genus *Dendrobates,* females have been seen to carry the tadpoles in the same manner as the males.

TWO-TONED ARROW POISON FROG
Phyllobates bicolor

The Two-toned Arrow Poison Frog is distributed throughout the deciduous woodlands of tropical South America. This tree frog climbs dexterously on the leaves of trees or bushes, catching small insects and their larvae, and small spiders. In the breeding season, the female is accompanied by the male as she lays about 10 relatively large eggs. The male then guards them until they hatch. The tadpoles crawl on to his back and attach themselves firmly. The male soaks them regularly in tiny pools of water in leaves or tree hollows. After 2 weeks the tadpoles leave their father's back and settle for 5 — 6 weeks in a pool of water. Here they grow legs, their tails are absorbed, and the tadpoles change into tiny frogs which take up residence in the trees.

Three-striped Arrow Poison Frog

Two-toned Arrow Poison Frog

Like the previous species, it is used to produce arrow-poison.

BLOMBERG'S TOAD
Bufo blombergi

Blomberg's Toad is the largest of the toads, and the second largest anuran in the world after the African Giant Frog. It reaches a length of over 25 cm not including the legs, and a weight of up to 1.5 kg. Despite its gigantic size, scientists discovered this amphibian only recently. It was first caught in 1951 by the Swede Blomberg, in Colombia, and it was also found later in Ecuador. This toad lives near slow-running water or on the banks of pools and lakes. It is nocturnal, seeking its food after sunset. In dense forests it leaves its shelter during the day when it is raining. It hunts spiders, large insects and larvae, and mature specimens devour small vertebrates, such as young rodents, lizards and snakes. The prey is swallowed whole, and an adult toad can eat as many as four mice one after the other.
In the breeding season toads from over a wide neighbourhood congregate in shallow pools, and the females lay several thousands of eggs in gelatinous strings. These hatch into tadpoles which live in the water for 2 months before completing their metamorphosis into small toad.
The South American Indians have always known of the existence of this toad. They use it to make poison for their arrows. Its skin glands contain a large quantity of strong poison which quickly kills small animals. Carnivores learn to avoid these toads as food because the poison causes muscular spasms. The venom has an effect similar to that of digitalin extracted from foxgloves. This toad can normally be held in the hand as its poison only affects the skin of particularly sensitive people. However, acute eye inflammations can result from contact with this poison.

Blomberg's Toad

CUBAN TREE FROG
Hyla septentrionalis

The Cuban Tree Frog is one of the most abundant and also one of the largest species of tree frogs. It is found in Cuba and the Bahama Islands, and also in south-east Florida, where it was introduced among banana clusters in which it sometimes shelters. The female of this huge tree frog measures up to 14 cm in length, the male being smaller. The Cuban Tree Frog dwells near water in forests, parks and gardens, living in trees or on rocks and walls. The frogs enter water only in the breeding season, when the female lays over a thousand eggs in clusters. By day, tree frogs shelter in damp holes or beneath peeling bark, often many of them being found close together. They are able to store water in their bodies and do not suffer from desiccation in periods of drought. After sunset, they begin to feed, climbing tree trunks,

branches and walls as they search for insects and spiders. Larger specimens catch small anoline lizards. In the mating season, the males produce loud hoarse cries, and as the frogs congregate in large numbers at spawning time, the noise is considerable. Like other tree frogs, this species can change colour when excited or when there is a change, as for example in temperature, in the external environment. Its colour ranges from whitish to spotted brown.
Tadpoles hatch out after a week and feed on tiny crustaceans, worms and insect larvae. They change into smallish frogs within two months and then leave the water and grow to adult size on land.

Cuban Tree Frog

BLACKSMITH TREE FROG
Hyla faber

The Blacksmith Tree Frog reaches a length of 9 cm, and lives in southern Brazil and Argentina. It is found in trees in forests, parks and gardens, near shallow water. It gets its name from its characteristic high voice, like the sound of a hammer hitting an anvil. These loud sounds are made by the male in the breeding season. At this time he prepares a special nest for his offspring, on the edge of a muddy swamp or in a large shallow pool. Suitable places are always occupied by a number of males. They dig up the mud, heaping it around themselves to form circular walls about 10 cm high and enclosing an area of water about 34 cm across. The insides of the walls are carefully smoothed with the forelimbs, to keep the water in. At night, each male settles in his pond and calls until he attracts a female. After mating, the female lays 50 — 100 eggs in the pool. Here the developing tadpoles are protected from predatory fishes and water insects. Only a few tadpoles usually survive, because there is only a limited amount of food, mainly insects, crustaceans and worms, in the small pond. However, a sudden downpour often floods the pools and releases the tadpoles. Over the next few weeks the tadpoles change into frogs which leave the water and take up residence in the trees.

LEAF FROG
Phyllomedusa hypochondrialis

The leaf frog *Phyllomedusa hypochondrialis* is widespread from Paraguay and Bolivia north to Guyana and eastern Brazil. It measures only about 4 cm and has slit-like eye pupils. It lives on tree trunks and leaves in forests and bush-covered areas beside water. In the breeding season, each pair finds a broad leaf, about 60 cm above the surface of the water. With their hind legs they fold the leaf from its tip rolling it into the shape of a funnel and glueing the edges together with a sticky secretion. The female lays a few eggs inside the leaf and the male fertilizes them. Then they roll the leaf up a little more and the

Blacksmith Tree Frog

Leaf frog
Phyllomedusa hypochondrialis

female again lays a few eggs. In this way they continue up to the stem. The eggs in the folds are protected from predators, and as the leaf tube remains open at both ends, the newly hatched tadpoles fall down into the water where they develop into frogs. This frog feeds on insects, larvae and small spiders. A further 25 species of this genus are distributed in an area from Central America to Argentina.

TREE FROG
Agalychnis callidryas

The tree frog *Agalychnis callidryas* is beautifully coloured and measures about 8 cm. It dwells in the forests ranging from the state of Vera Cruz in Mexico south to Guatemala. During the day this frog sleeps with half-open eyes, attached to the underside of a broad leaf. It sets out at twilight to hunt for spiders, insects and their larvae, and the occasional tree mollusc. Among the greenery its red eyes shine like huge rubies. At the end of the dry season the frogs form pairs. After mating the female lays her eggs on leaves overhanging the water. With the coming of rain the eggs are washed into the water, and the tadpoles develop in a few days. By spending a shorter time in the water, the eggs are less likely to be eaten by fishes and reptiles.

TREE FROG
Pachymedusa dacnicolor

The tree frog *Pachymedusa dacnicolor* is one of the larger species, reaching a length of 12 cm. It is found along the Pacific coast of Mexico, from the south of the state of Sonora to the Isthmus of Tehuantepec. Unlike other related species, it is predominantly terrestrial. It feeds on insects, spiders and other tiny invertebrates, and also catches young iguanas. Sometimes, especially in the breeding season, it climbs into the trees. There it lays its eggs on leaves in the same way as *Agalychnis callidryas*. In the mating sea-

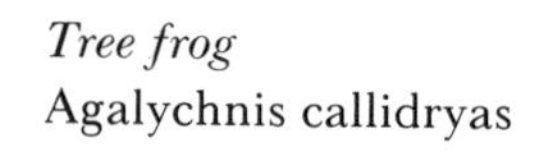

Tree frog
Agalychnis callidryas

Tree frog
Pachymedusa dacnicolor

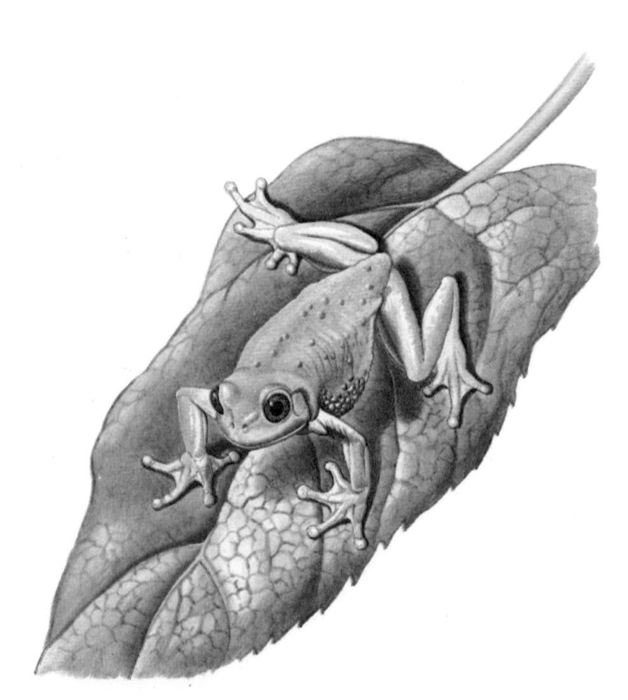

son, the male produces a far-carrying, drum-like call. This tree frog is nocturnal. When caught in a beam of light in the darkness, its eyes shine like gold and are visible over a distance of 50 m or more.

Darwin's Frog

DARWIN'S FROG
Rhinoderma darwini

Darwin's Frog is a tiny amphibian, reaching a length of only 3 cm. It inhabits coastal woodlands of southern Argentina and Chile where it occurs in many colour varieties, usually yellow, red and brown, but turning a uniform buff colour in autumn. The male has an inflatable throat sac, which intensifies the clear sounds he delivers in the breeding season. The female lays 20 — 30 eggs in damp leaves on the ground, while several males gather around her and guard the clutch until the larvae hatch, which is usually after 2 — 3 weeks. Each male then takes 6 — 8 larvae and puts them in his throat sac. Here they undergo metamorphosis and leave their shelter as fully-formed frogs. They disperse, and like their parents they hide in damp fallen leaves where their coloration enables them to escape attention. They feed on tiny insects, their larvae, worms, spiders and molluscs. Despite its perfect protective coloration, Darwin's Frog often falls victim to snakes and birds.

MARSUPIAL FROG
Gastrotheca marsupiata

The Marsupial Frog inhabits the forests of Ecuador and Peru. The female reaches a length of 8 cm, while the male is about a third smaller. This frog frequents mountain slopes where torrents of water pour down or become absorbed by moss and soft soil. There are no quiet pools in which its eggs can develop. In order to keep them moist, the female carries them on her body until they hatch. In the breeding season, the male grasps the female with his front limbs and after mating remains lying on her back. After about 24 hours, the female begins to lay eggs continuously for about 45 — 60 minutes, sometimes producing as many as 200. The male catches them and inserts them with his front limbs into a brood pouch on the lower part of the female's back. When it is full, this pocket is closed by a fold of skin, and here the eggs develop for 6 — 7 weeks. After hatching, the tadpoles stay for a short time in the pouch feeding on their yolk sacs. At a certain stage of their development, the female goes into the water, usually at night or early in the morning, and loosens the opening of the sac with the longest toe of her hind limb. The tadpoles swim out and complete their metamorphosis into frogs in water.

The Marsupial Frog is predominantly terrestrial, although it occasionally climbs on to the leaves of trees and shrubs. By day it shelters in underground holes, from which the male's mating call can be heard. It seeks its food after twilight, hunting insects, spiders and worms.

Marsupial Frog

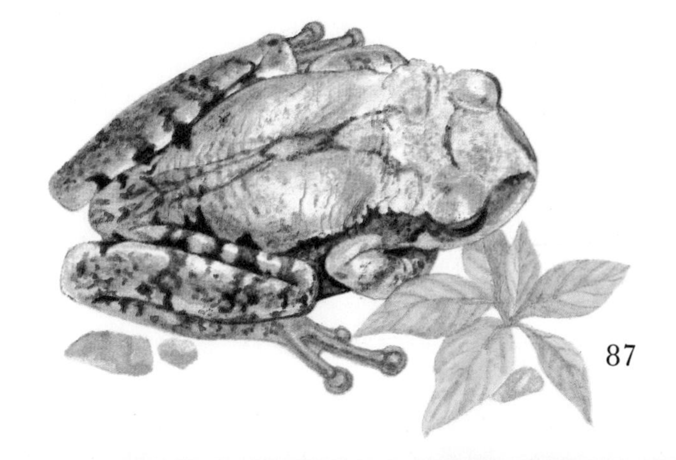

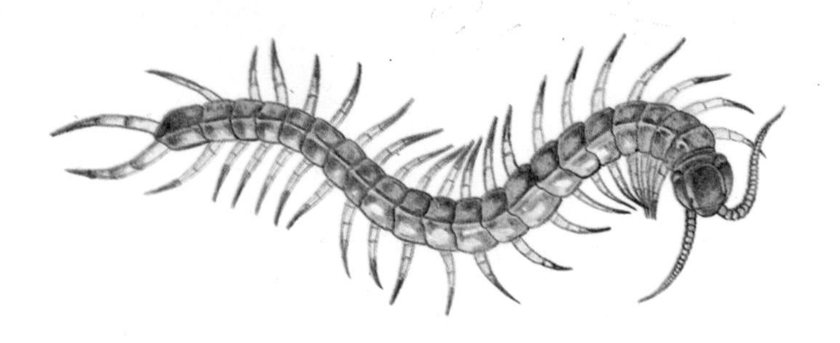

Centipede Scolopendra viridis

CENTIPEDE
Scolopendra viridis

The centipede *Scolopendra viridis* is widespread in wooded areas from Brazil northwards across Central America to the southern states of the United States. It also occurs in parks and gardens. This invertebrate measures up to 13 cm in length and as an adult has 21 pairs of legs. It frequents damp localities, sheltering under flat stones, trunks or peeling bark during the day. At night it forages in crevices and holes, searching for insects, worms and molluscs. Large centipedes feed on tiny iguanas or frogs. These large tropical centipedes pounce on their prey, seizing it with pincer-like mandibles, which contain poison glands. The poison is effective on small animals, but is not dangerous to larger ones. Its bite is painful for man and may give rise to swellings lasting several days in sensitive people. However, centipedes are not aggressive and bite man only in self-defence.
The female lays 50 — 100 eggs on the ground, among heaps of fallen leaves or in a large hollow under a stone, and envelopes them with her long, flat body to protect them from predators. The hatched young have the same number of segments as the adults, but fewer legs. They disperse and their mother takes no further care of them. The young undergo a number of moults, growing another pair of legs each time. They mature at 3 years of age. They are solitary, forming pairs for only a short time in the breeding season.

Millipede Orthoporus pontis

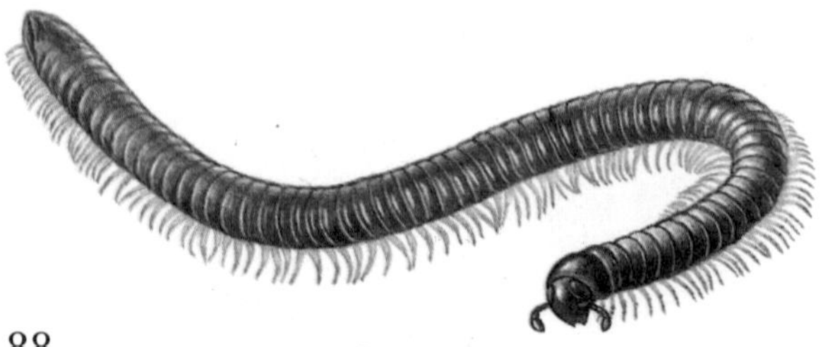

MILLIPEDE
Orthoporus pontis

The millipede *Orthoporus pontis* inhabits humid woodlands in northern Mexico, and is quite common in western Texas. It is one of the larger species, reaching a length of 12 cm. The dorsal plates of this invertebrate are reinforced with calcium carbonate, making them extraordinarily tough and durable. The millipede lives under fallen tree trunks, leaves or stones, leaving its shelter only at night when it seeks food. It feeds solely on plant debris, such as decaying leaves, or soft fallen fruits. It is a harmless creature, falling victim to a variety of predators. Its chief defence mechanism is a foul-smelling caustic fluid sprayed from glands along its body, which poisons small animals and repels some mammals. When in danger, this millipede can curl into a ball, and its carapace sometimes saves it.
The female constructs a special case from a mixture of clay and excrement stuck together with a secretion from her anal glands. She lays eggs into this chamber, and then walls them in. The case has an aerating shaft. The female conceals the structure with pieces of leaf, and takes no further care of her offspring. The young have only a few body segments and three pairs of legs. New segments and legs appear at each moult.

BIRD-EATING SPIDER
Avicularia avicularia

The bird-eating spider *Avicularia avicularia* often frightens people by its gigantic size, hairy body and legs, and enormous mandibles. It is found in the primary forests and plantations of tropical South America. It reaches a length of 5 — 6 cm, but its long legs make it look much bigger. In spite of its terrifying appearance, it is harmless to man. It is not aggressive and bites only in self-defence. Furthermore, it only produces a very weak poison, which is not fatal even to small animals. This spider overpowers its prey by sheer strength, pouncing on the animal in a huge leap and killing it with its strong jaws. It mainly hunts insects and their larvae, but adults also catch small geckos, anoline lizards, frogs, and small birds. The spider injects digestive juices into the prey to decompose the soft parts of the body, and then it sucks up the resulting liquid. Only the tough remains are left, such as skin, scales, bones and tendons, or the hard chitinous parts of insects, such as wing-cases and legs. The bird-eating spider is nocturnal, hunting after sunset, and sheltering in holes among roots, or under stones, during the daytime. It weaves a tubular retreat, closed by a web lid equipped with a curtain which can be fastened, and no predator can get in.

Inside the finely lined web tube the female lays 100 — 200 eggs in a cocoon. The young hatch after 25 days but stay in the cocoon for another 20 days, before biting their way out. The female holds the cocoon between her front legs to guard it. When the weather is hot and dry, she carries the cocoon outside and holds it above a pool of water. The cocoon is moistened by evaporation, and the eggs do not dry up. From the time the eggs are laid until the hatching of the young, the female takes no food. The young moult for the first time in the cocoon, and for a second time a few days after leaving it. When shedding their skins, these spiders remain motionless for 12 — 16 hours, and are highly vulnerable to attack.

Outside the mating season, encounters, especially between males, often result in fights. The spiders stand on their hind legs, and raise their front legs to attack each other, wounding and sometimes killing their adversaries.

These aviculariids often reach North America and Europe in imported bananas. They have to be handled carefully, because their fine, broken hairs can cause eye inflammations.

About 500 aviculariids live in South America, but none of them reaches the size of *Avicularia avicularia.*

TROPICAL GOLDEN WEB SPIDER
Nephila clavipes

The Tropical Golden Web Spider is one of the most interesting spiders of the South and Central Ameri-

Bird-eating spider Avicularia avicularia

Tropical Golden Web Spider

can tropics. The female measures 25 mm, while the male measures only 4 mm. This spider has conspicuous tufts of hair on its legs. It builds huge webs, over 1 m across, between tree trunks. These webs are so strong that they often trap small birds, and South American Indians use them to catch fish. This spider preys particularly on butterflies, flies and other flying insects, caught in its snare. Young spiders spin the whole web, while the adults construct only the centre part and any irregular projections.
Young spiders hatch out of a cocoon and after a few days they disperse. During their life they moult twelve times, which is almost double the number of moults undergone by other species of spiders.

Tree-hopper
Bocydium globulare

Lantern-fly Fulgora lanternaria

Leaf-cutting ant Atta sexdens

TREE-HOPPER
Bocydium globulare

The tree-hopper *Bocydium globulare,* is a mere 5 mm long. It is a curious insect found in South America, and it is very common in some localities. It grows a very strong protuberance which projects from its front covering and turns back to reach as far as the end of the abdomen. This protuberance bears four globular, stemmed formations covered with fine hairs. The significance of these strange formations remains unknown.
This tree-hopper lives on bushes in woodlands and plantations where it sucks sap from thin twigs on the ends of branches. In the breeding season, the female cuts into a thin twig with her ovipositor and lays eggs in the opening. The bark then seals over, shutting in the eggs. If a plant is attacked by a large number of tree-hoppers, the terminal twigs dry up. The hatched larvae, called nymphs, also feed on plant sap. They differ from the adults in having no protuberances.

LANTERN-FLY
Fulgora lanternaria

The lantern-fly *Fulgora lanternaria* is common in the tropical rainforests of South America. This large bug is 9 cm long, spans about 13 cm, and has an enlarged peanut-shaped head. During the day it rests with folded wings, pressed to a tree trunk. Its first pair of wings is cryptically coloured, enabling the insect to escape attention, but when disturbed, it suddenly opens its wings to reveal two large patches of colour on the lower wings. These spots look like eyes, and they deter animals both large and small. Nevertheless, lantern-flies readily fall prey to birds and iguanas. These insects are active after sunset, flying from tree to tree or among the bushes, to settle on branches and suck the sap.
The larvae also have large heads. They live in soft soil beneath the trees, and move by means of their front legs, which are adapted for digging. Here they search for delicate roots, and suck the juice. When they are fully grown, they climb up the tree trunks and attach themselves to the bark. Very soon the skin on their backs bursts, and the adult insects crawl out. The lantern-fly does not emit light, and it is not a fly.

In an early description it was wrongly said to be luminous, and so it obtained its name. The purpose of the strange head formation is unknown, but scientists believe it serves some function during the search for a mate in the breeding season.

LEAF-CUTTING ANT
Atta sexdens

The leaf-cuting ant *Atta sexdens* lives in woodland regions of tropical South America. It is only 2 — 4 mm long, but it is one of the most serious pests of deciduous trees and bushes. These ants use their powerful mandibles to cut pieces of leaf from all species of deciduous trees. Then they carry their spoils to their underground nests. Since the colonies are very large, the inhabitants of just one anthill can cut off thousands of leaves in a single night, denuding even the largest trees. They set off on long, nocturnal foraging trips in their millions, and cause enormous damage, particularly in coffee plantations. This pest is to some extent controlled by the gassing of its underground nests.
The ants carry their loads of leaves into the nest and process them by chewing them into pulp, which is wetted by saliva and excrement. From this material, they build 'gardens' about the size of a football, and here they grow fungi of the genera *Rhozites* and *Tyridiomyces,* carefully removing all other fungi. In the warm, humid environment, the fungi grow quickly, but the ants continually restrict their growth by biting off the ends. The regrowth is called 'ambrosia', and is the material on which the ants feed. Each species of leaf-cutting ant cultivates a different species of fungus. When swarming takes place, each queen carries spores of the fungus for the new nest in a specially adapted mouth cavity.
The queen lays her eggs in an underground chamber. Here the hatched larvae are fed on other eggs, masticated by the queen, and eggs are also added to the base of the fungus garden. The queen, too, feeds on her own eggs, for she is unable to take any other food. The fungus, fertilized by her excrement, takes some time to become productive. After a few weeks the larvae pupate, and within a few days female workers emerge, and begin to build a new underground colony. They expand the garden, collecting and chewing ever more leaves as the colony grows.

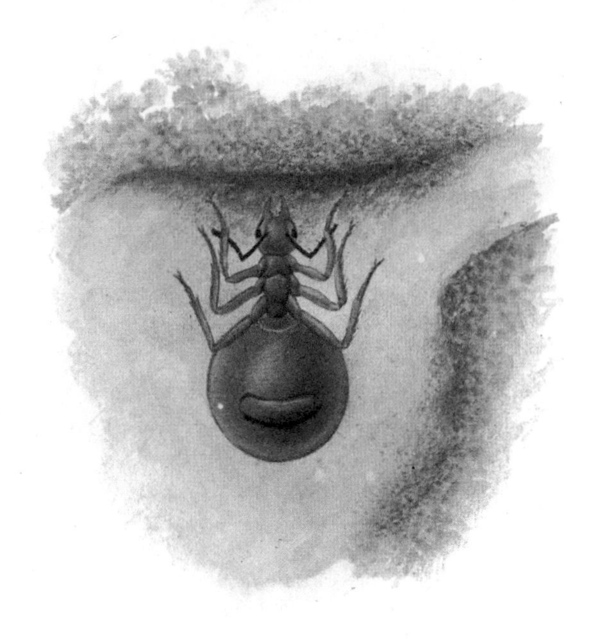

Honey ant
Myrmecocystus hortideorum

HONEY ANT
Myrmecocystus hortideorum

The honey ant *Myrmecocystus hortideorum* inhabits the edges of South American tropical forests. Colonies of these ants build underground chambers, the largest of these being perfectly clean and smooth inside. From its ceiling hang clusters of pea-sized honey-pots. These are female ants, the enlarged abdomens of which are filled with honey. These live containers are incapable of movement, because all their other organs are pressed aside by their enlarging abdomens. Honey is continually fed to them by female workers, which collect it from flowers and regurgitate it into the mouths of the strangely transformed females. The workers both feed and clean the honey-pots. When they are full, the colony is able to survive through the long rainy season, or any other period of bad weather, without being short of food. The workers signal to the honey-pots with their antennae when they want them to discharge honey. Local Indians search for the nests of these ants, dig them out, and take the honey.

Scarab beetle
Enema pan

SCARAB BEETLE
Enema pan

The scarab beetle *Enema pan* is unusual in appearance. The male reaches a length of 6 cm and has a huge, forked horn growing from its carapace. This curious creature inhabits the vast tropical forests of South America. It is very common in some areas and can be seen during the day as it crosses the forest paths. However it is mostly active at night. The female lays her eggs in the rotting wood of trunks or stumps of deciduous trees. The larvae have fat bodies and are about 10 cm long. They have strong mouthparts and feed on decayed wood. After 3 years they pupate in a cocoon made of clay and wood pulp. The male pupae already have visible horns. The hatched beetle remains in the cocoon for several days or even weeks, before its wing-cases and the other sclerotic parts of its body harden. Then it perforates the wall of the cocoon and crawls out.

Jewel beetle
Euchroma gigantea

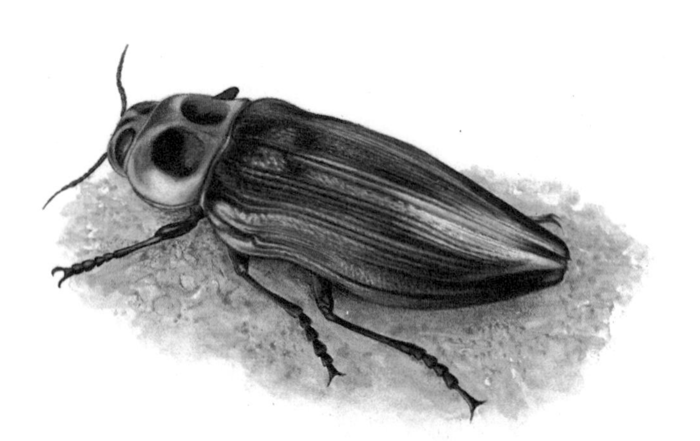

JEWEL BEETLE
Euchroma gigantea

The jewel beetle *Euchroma gigantea* is one of the largest and most beautiful of the beetles. It is about 7 cm long and 2.5 cm wide and its wing-cases have a metallic sheen. It inhabits forests from Mexico to Brazil, occurring in several subspecies differing in size and coloration.
Euchroma gigantea often rests on tree trunks, displaying its colours in the sun. Local Indians catch these beetles and make necklaces and bracelets from their wing-cases. Sometimes they use whole beetles, for the underside is bright red and also has a metallic sheen. The female lays her eggs in the bark of trees, and the larvae develop in the trunks. They bore tunnels in the wood, and pupate there, the adult beetles crawling out of the tunnels after emerging from the pupae.

BEETLE
Chiasognathus granti

The beetle *Chiasognathus granti* is common in the woodlands of Chile, especially in the mountains. The male measures about 8 cm and has very long, pincer-like, saw-toothed mandibles. The female has short mandibles and is also smaller. These beetles have conspicuously elongated front legs.
They live in deciduous trees, feeding on sap which flows when the bark is damaged. The female lays her eggs in decaying trunks and stumps. The developing larvae feed on rotting wood and reach a length of 10 cm. The larvae stage lasts for several years. The sexes can be distinguished in the pupal stage, for the

Beetle Chiasognathus granti

males have large visible mandibles. *Chiasognathus granti* is a nocturnal beetle, flying after sunset around the edges of forests and clearings.

BEETLE
Dynastes hercules

The beetle *Dynastes hercules* is one of the largest of the beetles, the male reaching a length of 17 cm. It has a massive, elongated carapace terminating in a long horn. From below the head grows another, shorter, dented, upward-turned horn. These huge protuberances are not mandibles, but serve as an efficient means of defence, for they can be manipulated like pincers. The female has no horn and measures only 9 cm.
Dynastes hercules inhabits tropical regions of Central America, particularly the Antilles. It lives in deciduous woodland, where the female lays her eggs in rotting stumps or fallen tree trunks. The developing larvae measure up to 12 cm, and the horns are visible in the pupae.

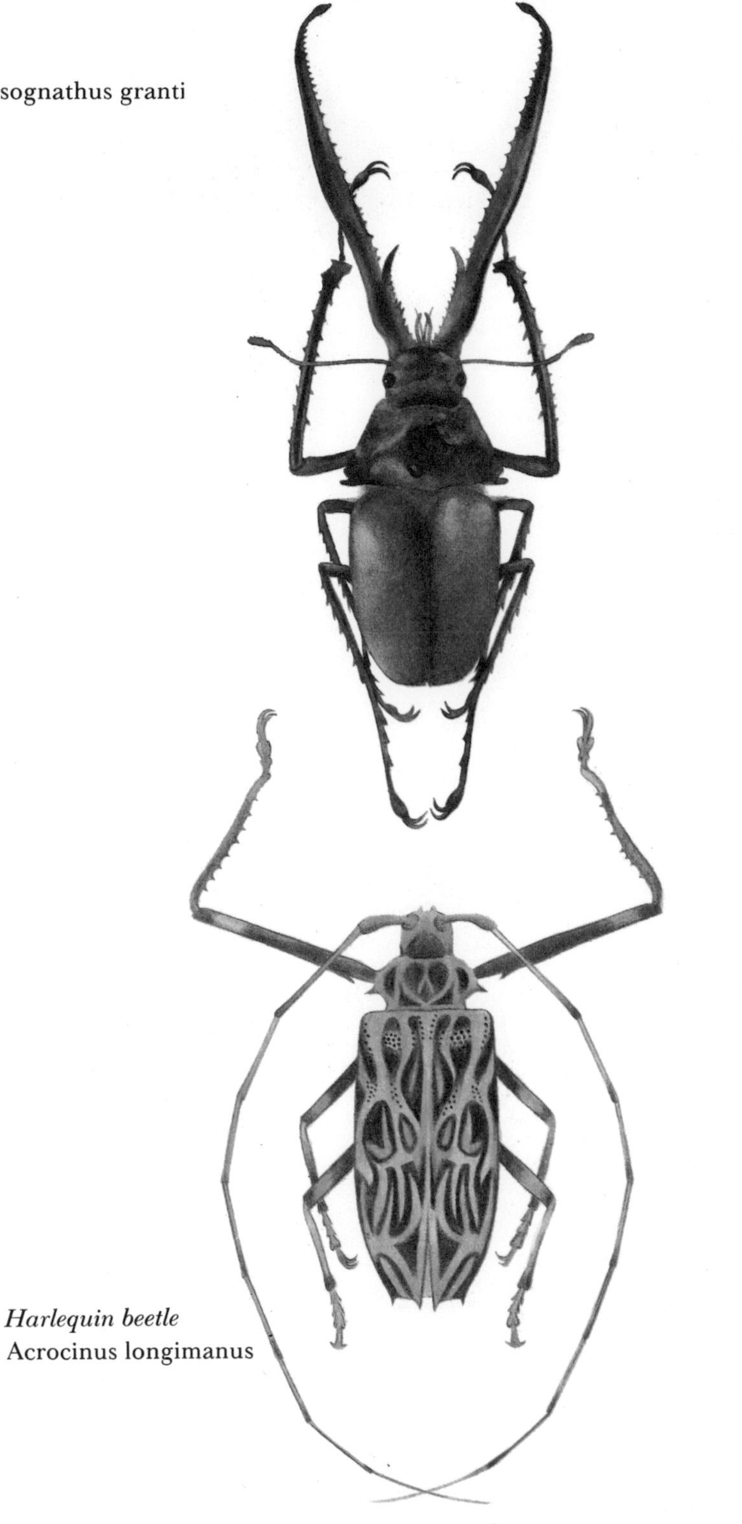

HARLEQUIN BEETLE
Acrocinus longimanus

The harlequin beetle *Acrocinus longimanus* is distributed from Central America to southern Brazil, where it is a pest of citrus plantations. Its body length is about 8 cm, but overall it is a large beetle, for it has remarkably long front legs, measuring up to 17 cm. The antennae are also longer than the body. This beetle hides during the day under peeling bark, and becomes active after sunset. The female lays her eggs in the bark of fig or citrus trees. The larvae develop in the wood and reach a length of 8 cm. A tree attacked by large numbers of larvae becomes riddled with tunnels and eventually dies.

Harlequin beetle Acrocinus longimanus

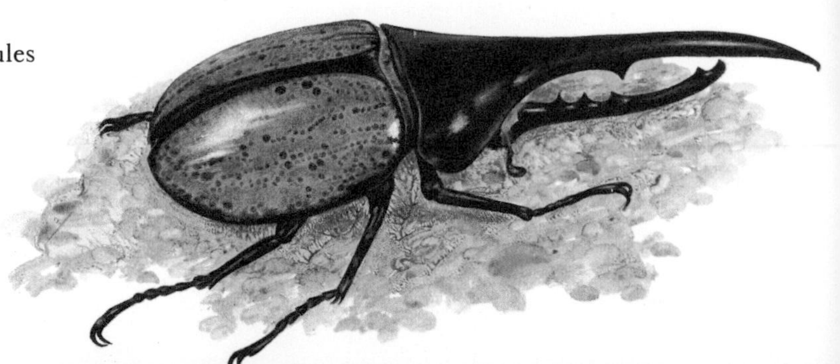

Beetle Dynastes hercules

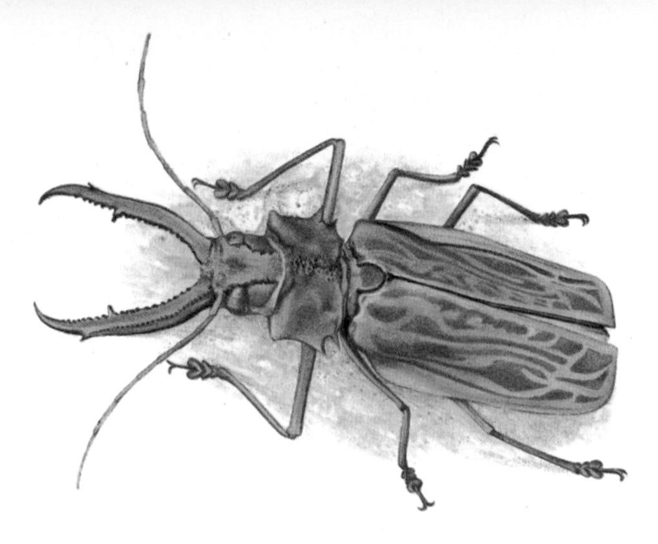

Beetle Macrodontia cervicornis

BEETLE
Macrodontia cervicornis

The beetle *Macrodontia cervicornis* is another giant of the insect realm. It lives in the forests of Brazil, particularly in the south. It has huge mandibles, 4 cm long and 7 mm wide at the base, with dented protuberances on the inside. The antennae are rather short, reaching a length of only 6 cm. The female lays her eggs in the bark of deciduous trees, and the larvae bore tunnels into the wood. The larvae grow to a length of 15 cm, and make cocoons of wood pulp before pupating. The native Indians chip out the huge fleshy larvae and eat them.

MOTH
Castnia icarus

The moth *Castnia icarus* has a wingspan of about 10 cm. It has markedly long antennae, and the forewings are greenish with a metallic sheen. It is a fast and expert flier, constantly fluttering in the sunshine above the tallest treetops, and only descending in the early morning to look for pools from which to suck water. Its native land is the primary forests of Brazil and Guyana. The female lays her eggs in crevices in the bark of trees and the caterpillars bore into the trunks. They often drill corridors inside the branches of trees or bushes, cutting holes to let themselves out before pupation.
South America is the homeland of other moths of the genus *Castnia,* some with wingspans averaging 18 cm. The larger species defend their territories, driving away both insects and small birds by flying up sharply and forcing them to retreat.

BUTTERFLY
Pereute leucodrosime

The butterfly *Pereute leucodrosime* lives along the edges of forests and bushland of tropical South America. Its wingspan averages 7 cm, and the butterfly is

Moth
Castnia icarus

Butterfly
Pereute leucodrosime

Zebra butterfly

very colourful, often entirely lacking the white coloration which gives it the name 'white butterfly' in some languages. Early in the morning, it flies down to the ground to seek water in damp places, but during the day, it flies above the treetops and tall bushes. Large numbers of these butterflies may sometimes gather in a favourable spot.

ZEBRA BUTTERFLY
Heliconius charitonius

The Zebra butterfly has a wingspan of about 8 cm, and inhabits woodland, and river banks overgrown with vegetation, in tropical South and Central America. It often visits southern California and Kansas, a journey which takes several weeks, for it is a slow, weak flier. The reason for these long, northbound trips remains something of a mystery. This butterfly produces a characteristic scent, similar to that of certain bugs. The scent is strong as far as ten metres away and has the effect of repelling many animals which would otherwise eat it.

During the day the Zebra flies above the trees and bushes, but it settles before sunset, at a regular site to which it is guided by its sense of smell. Colonies of these butterflies spend the night together and disperse in the morning.

The female lays her eggs on climbing passion-flowers and the caterpillars devour the leaves. The caterpillars are blue-grey, with tall thorny tufts and two small horns behind the head. They live solitarily and attach themselves to leaves before pupation. The pupa is covered with protective hooked spines and hangs head downwards. Males are attracted by the scent of female pupae before the females emerge. They can smell them at a distance of over 100 m, which is an incredible achievement.

Butterfly
Morpho hecuba

BUTTERFLY
Morpho hecuba

The butterfly *Morpho hecuba* is the largest diurnal American butterfly, with a wingspan of up to 15 cm. The female is usually larger than the male. Unlike its relatives, this species is orange-brown to brownish. The underside of its wings is also brown, but is covered with large orange spots with black rims and having white dots in the middle. *Morpho hecuba* inhabits the region between the Amazon and the Orinoco in Brazil, where it lives mainly on the edges of forests or along wooded river banks. It has excellent powers of flight and moves high above the treetops. It often flies above the rivers, and may travel many kilometres before returning in the evening. In the early morning, these huge butterflies alight to suck water from pools and marshes. This habit is made use of by collectors, who catch the beautiful butterflies at this time, sometimes luring them by means of huge mirrors, which the butterflies mistake for glittering pools of water.

Butterfly
Morpho anaxabia

BUTTERFLY
Morpho anaxabia

The butterfly *Morpho anaxabia* is one of the many species of this genus. Its iridescent blue wings span about 14 cm. The remarkably high gloss of the wings is caused not by pigment, but by the refraction of the sun's rays in the hollow scales which cover the membranous wings like tiles on a roof. This butterfly inhabits the forests of Brazil, especially those near rivers. During the day, it flies high above the treetops, descending in the morning to drink. It is always very wary, and is not easily caught.

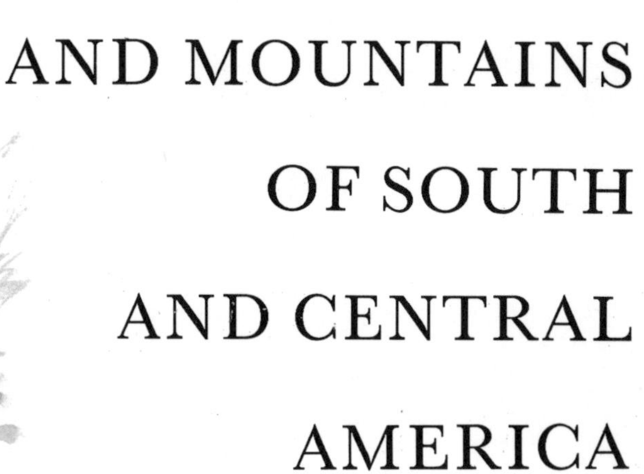

THE DESERTS, SAVANNAHS AND MOUNTAINS OF SOUTH AND CENTRAL AMERICA

We are now going to leave the primary rainforests, and cross immense plains to reach one of the most massive mountain systems of the world, the Andes. A large part of Central and South America is covered by vast tracts of desert and semidesert. They are found in Mexico, along the western coast of South America in Peru and Chile, and in southern and south-western Argentina. The stone-strewn, sandy arid landscape is sparsely relieved by low brushwood, but here the cactus is king, being found in countless forms from miniature plants to giants, tens or even hundreds of years old. Some cacti of the genus *Cereus* seem to belong to a fantasy world, with their huge pink, red, or yellow flowers, drawing one's attention from a distance of several kilometres. In Mexico, one can see the world's largest cactus, *Carnegiea gigantea,* which sometimes towers to a height of twenty metres — taller than a six-storey building. Birds often perch on these cacti, to await their prey. Woodpeckers drill holes in them, and after they have left their nests, others take up residence in their abandoned homes. The world of sand and stones is far from deserted.

In the stony regions of Mexico and Central America many species of agave are to be seen, their sharp-tipped leaves forming rosettes, up to 2 m high. Agaves are now cultivated in plantations in Central and South America, the crushed leaves yielding long, tough fibres called sisal, used in the manufacture of ropes and various other products.

There are other kinds of plant life to be seen here. Where patches of soil have been protected by stones, there are various species of low, shrub-like or tree-like euphorbias. The dry semideserts of Patagonia at the southern tip of Argentina are covered by short, dry grassland, containing occasional low, crooked trees of the genus *Nothofagus.* In these southernmost regions, rare falls of winter snow supply much-needed moisture to the arid soil.

Only the desert stretching away north from the city of Lima in Peru is empty of life. There is no vegetation at all, and animals seldom stray there.

In contrast, life teems in the savannahs. These are vast open grasslands, characteristic of some regions of Central and South America. They are called by different names depending on their geographical situation and predominant flora. The best-known is the Argentinian pampas, which stretches as far as Bolivia. These grasslands cover low-lying areas in the tropics and subtropics, and are home to many kinds of animals which enjoy their warm climate. This area is well known for cattle breeding. There are fringes of woodland along the rivers, but generally the pampas features solitary groups of trees, the homes of predatory birds which hunt the small rodents of the grasslands.

Savannahs in the south-west of Brazil are called campos. They are dotted with scattered trees and areas of sparse woodland. Campos occupy an area of more than 250 000 square kilometres — an area greater than that of West Germany or Great Britain. Like the tropical rainforests, they are evergreen but they lack their wild luxuriant character. The broad-topped trees are only up to 2 m high and have few branches. They do not have to fight for their place in the sun by growing straight up like the trees of the rainforests. The most frequent species are mimosas, especially *Mimosa pudica,* and myrtle trees and shrubs.

Savannahs could well be called 'the land of rodents', for these creatures abound in the grasslands. They in turn attract carnivorous animals, and predatory birds enjoy ample supplies of food as well as ideal nesting conditions. The dry, favourable climate which prevails in winter gives way to regular summer rains which flood vast areas. Then the rodents seek refuge on patches of higher ground, where they easily fall prey to a variety of carnivores, including the jaguar, king of the forest. Jaguars come out even in daylight, to prowl in the shallow floodwater and take their pick from the rodents and other small mammals stranded there.

The northern provinces of Venezuela are covered by extensive savannahs called llanos.

Here the grassy cover is taller and more continuous. Trees are few and far between though palms of the genera *Mauritia* and *Corypha* grow in moister localities along the coast. But the Orinoco, 'Father of Waters', treats Nature's children cruelly. Every year, from November to June, it floods hundreds of square kilometres, drowning not only underground animals but arboreal species as well, for the water loosens the roots of trees growing near the river, and whirls the trees away complete with the monkeys, sloths, snakes and other animals which occupy them.

Savannahs are also found along the western coast of Central America and Mexico, and on some islands of the Antilles, particularly Cuba. The vegetation here consists of palms, myrtle trees and shrubs with edible fruits, and low mimosas. There are also many varieties of cactus, characteristic of this locality, some growing to considerable heights and having bizarre shapes.

Arid regions are the haunt of many terrestrial and arboreal iguanas and colourful anoline lizards. Other inhabitants include various species of snakes, such as grass-snakes of the genus *Dromicus,* small boas of the genus *Tropidophis,* which only reach a length of 1 metre, and other generally harmless reptiles. However, in Mexico one can encounter the most venomous of all reptiles, the rattlesnake. The plains also offer a magnificent spectacle of colourful butterflies, and there are many species of beetles, ants, wasps and bees.

Having crossed the savannahs and deserts we are now facing the steep Andes which tower over the western coast of South America running parallel to it for 8 000 km from the Caribbean to Tierra del Fuego. The circulation of warm air in the mountains makes it possible both for people and animals to live several thousand metres above sea level. Evergreen bushes of the genera *Baccharis, Colletia* and *Acacia,* as well as various cushion-like plants, such as *Saxifraga,* flourish up to 3 500 m.

In winter, despite the bitter cold and strong winds, plants and animals are still found up to 5 000 metres, and Indians of various tribes including the Aymara and Chibcha dwell here. In the extreme conditions under which they live, they rely heavily on one of the most useful animals of the continent — the llama — ancient symbol of life in Indian culture and art. Both domesticated llamas and alpacas provide milk, meat, wool and hide, and the Indians make use of even their fat, bones and droppings. Herds of wild guanacos can sometimes be spotted on the rocky slopes, but these llamas are becoming rare and are threatened by extinction.

While we are here we must not miss the legendary Galápagos Islands. These lie in the Pacific Ocean, off the coast of Ecuador, and are one of the most remarkable areas of equatorial America. These rocky islands are of volcanic origin. Some parts are arid, but others are covered with grass, bushes, and occasional low trees. The fauna is very special and often unique, including the only species of iguana which seeks its food underwater. There are also large numbers of tortoises, from which the islands get their name. Here are 76 species of birds which exist nowhere else in the world. Experts still marvel at Darwin's finches, especially the genera *Geospiza* and *Camarhynchus.* Each of the thirteen local species differs markedly from the others, either in the shape of its beak, or by the kind of food it eats or even the places where it finds its food. Some of these species have acquired a skill unmatched in any other kind of bird. After uncovering insect larvae in crevices in the bark of trees by means of its stout bill, the Woodpecker Finch *(Camarhynchus pallidus)* uses cactus thorns to prise them out.

And now let us look more closely at the animals of the deserts, savannahs and mountains.

SPECTACLED BEAR
Tremarctos ornatus

The Spectacled Bear is the only bear of the South American continent. It is found in the equatorial region of the Andes from Venezuela to Peru and Chile, where it inhabits mountain forests at heights from 1 700 to 2 000 m. With a body length of 130 — 170 cm and a height at the shoulders of 50 — 70 cm, it is one of the smallest of the bears. It owes its name to the rings of light-coloured fur around its eyes. In many parts, this handsome carnivore has been hunted to extinction for its fur.

During the day, the Spectacled Bear shelters in caves, leaving its den only at dusk. It roams throughout its territory, which it marks by scratching bark from the trees and perfuming the bare wood with secretions from its scent glands. It is omnivorous, voraciously eating anything from small mammals, birds, eggs and reptiles to insects. It also consumes fruit and digs up roots. Like all bears, it seeks delicacies, such as the honey of wild bees found in the trees.

The breeding season begins in June or July, and the solitary males start looking for mates. After a gestation period of about 8 months, the female gives birth to a single cub, or occasionally to twins or triplets. The underdeveloped, blind, sparsely furred cub weighs 300 — 350 g at birth. It is born during the winter when the bears are in their dens for their winter sleep. Throughout this time the mother does not take food and lives off her subcutaneous fat. She holds the cub close against her to feed it and keep it warm. It opens its eyes after 22 — 40 days and takes its first solid food at the age of 90 — 120 days. The young bear cub is very playful. It leaves the den after 2.5 months and matures after 3 years. The mother looks after it for 2 years, by which time it is independent and seeks its own territory.

TAYRA
Eira barbara

The Tayra is a well-built marten-like carnivore, found in several subspecies in the tropical region from southern Mexico to northern Argentina. It reaches a length of 100 — 115 cm including the tail, which is 38 — 47 cm long. The female is usually larger than the male and weighs up to 4.5 kg. During the day, tayras shelter in rocky or bush-covered sites.

Spectacled Bear

They set out after sunset to hunt rodents and small birds, but they also catch iguanas and snakes. They often take birds' eggs and catch fishes or frogs in shallow water. In quiet localities they seek food even during the day.
In the mating season pairs form and live together for a few days, but for most of the year, tayras are solitary. After mating, the female begins her maternal duties, lining the den with hair and feathers, and after a gestation of 42 days (sometimes longer), she bears 2 — 4 blind young. These gain their sight after a month, and at the age of 5 — 6 weeks begin to feed on prey brought by their mother. The female suckles them for over 8 months.

Tayra

GRISON
Galictis vittata

The Grison, another marten-like carnivore, reaches a length of about 65 cm, including the 15 cm long tail. Adult specimens weigh 1.4 — 3.2 kg. The Grison ranges from southern Mexico southwards to Peru, frequenting bush-covered and stony localities with solitary trees. It lives alone, mainly on the ground, but it sometimes climbs on bushes or up small trees. During the day, it hides in tree hollows near the ground, in small caves, or under boulders. It leaves the shelter at dusk, swiftly running through the countryside looking for prey. It feeds on small rodents, birds and birds' eggs, and occasionally catches small snakes, lizards, large insects and molluscs.
In the mating season, the male seeks a female and stays briefly in her company. After a gestation period lasting about 42 days, the female gives birth to 2 — 4 young in a softly lined den. These are born blind, opening their eyes after 4 weeks. The female nurses them for 5 — 6 weeks, and they become independent at the age of 3 months. The Grison lives to be 7 years old. When in danger it sprays its enemy with a foul-smelling secretion from its anal gland. This often deters the attacker.

HOG-NOSED SKUNK
Conepatus leuconotus

The Hog-nosed Skunk, a 70 cm long carnivore of the weasel family, lives in a broad zone ranging from Arizona, New Mexico and Texas, in the extreme south of the United States across Central America to the northern part of South America. It is very common in warm localities, where it frequents open country overgrown with bushes and occasional trees and rocky sites. During the day, it hides in caves and holes among stones, coming out to forage at dusk. It combs its territory, digging out larvae and worms, hunting small rodents and birds, taking the eggs of ground-nesting birds, and occasionally catching reptiles. It also collects soft fruit on the ground. It sometimes visits villages and takes poultry. Though often detected by a watch dog, it always escapes unharmed, for the skunk repels attackers by spraying them with an offensive-smelling liquid from vents under its tail.
After a gestation period of 42 days the female gives birth to 3 — 5 young, born blind and naked, and gaining sight after 20 — 30 days. The young suckle for 6 — 8 weeks, but begin to eat prey brought to them when they are about 35 days old. At 10 months,

Grison

Hog-nosed Skunk

they leave their mother and fend for themselves, becoming mature a month later. Their average lifespan is 7 years, but in some localities they are hunted for their fur.

MANED WOLF
Chrysocyon brachyurus

The Maned Wolf is a resident of the vast pampas of South America, from Brazil to northern Argentina. This handsome carnivore inhabits low-lying land, preferably marshland containing networks or rivulets and lakes rimmed with lush vegetation. It has a slender, vertically compressed body and long legs, which enable it to push its way through dense growth. It lifts both legs on one side alternately with those on the other side, which helps the wolf to move through tall grass and reeds. Adult specimens reach a height of 75 cm and over 20 kg in weight. The Maned Wolf is a large carnivore, but it feeds on small prey, mostly agutis, small birds, especially nestlings, and invertebrates. It roams through the swamps and catches frogs, fishes and crustaceans in shallow water. It also eats sweet fruit. Near villages, it sometimes takes poultry, but does not attack larger

Maned Wolf

Argentine Wood Cat
or Geoffroy's Cat

domestic animals. It hunts during the day or at dusk, sleeping by night in a thicket, a hollow tree or a cave. In December or January the female gives birth to 1 — 5 pups after a gestation period lasting 62 — 66 days. The young weigh about 0.5 kg at birth and are blind. They obtain their sight after 12 days. At the age of one month, they leave the lair for the first time but stay in front of the entrance. At 3 months, they begin to hunt. Adult wolves make sharp, barking noises.

ARGENTINE WOOD CAT or GEOFFROY'S CAT
Felis geoffroyi

The Argentine Wood Cat or Geoffroy's Cat is the smallest South American cat. It reaches a total length of 60 — 80 cm, but 25 cm are occupied by the tail. It covers a large area from Brazil southwards, frequenting the edges of deserts and bushland with scattered trees around forests and in the mountains. It is a good climber, and lives mostly in the trees. When in danger, it lies close along a branch, its spotted coloration blending perfectly with its background. In the same way it lies in wait for its prey, pouncing when a small mammal or bird approaches. It also climbs through bushes and treetops in search of nests. During the day, it hides in the dense crowns or in holes in the trees, beginning to hunt after sunset, like most feline carnivores. It sometimes catches small lizards or snakes, and when food is in short supply, it feeds on insects and molluscs. It seeks water in which to bathe and swim, sometimes catching fishes. After a gestation period of 56 days, the female gives birth to 2 — 4 young in a hollow tree. The

Great Anteater

kittens are born blind, gaining sight after 9 days. They suckle for 4 months, but at the age of 1 month, they begin to nibble at the prey brought by their mother, who later teaches them to hunt. The male takes no part in family life. The Argentine Wood Cat is solitary except in the mating season. Its lifespan is 10 — 15 years, but the Indians hunt it for its fur.

GREAT ANTEATER
Myrmecophaga tridactyla

The Great Anteater lives in localities where termites abound, in arid bushland ranging from Guatemala south to Paraguay and Argentina. It is common in mountainous areas where the temperature at night falls to below 0° C. It is well-equipped for the cold, for it has long fur, and when asleep, it covers itself with its long bushy tail. This massive animal measures up to 2.3 m including the 1 m long tail, and an adult anteater weighs up to 50 kg. It has a narrow, elongated head and long, toothless jaws, which cannot be opened, and a very long, narrow tongue which it can extend over 50 cm. Immediately after sunrise, the anteater sets out to forage. The fingers of its hands have strong, pointed claws which bend backwards so that the animal walks on its knuckles. These powerful claws are used to dig out termitaria or anthills, and the anteater collects the disturbed insects in hundreds on its long sticky tongue, eating soldiers, workers and pupae. Its tough hide protects it from the soldier termites' mandibles, and its small eyes are protected by thick lids. The food is crushed in the stomach by friction of the muscular walls which makes pulp of the prey. When foraging for food, the anteater moves with its snout close to the ground, sniffing out anthills and termitaria. When replete, it often seeks water and stays in it for hours, for it is an enthusiastic bather. It is also a good swimmer, easily crossing large streams. It has few enemies in the wild, only jaguars daring to attack it. In this event, the anteater sits on its hindlegs and hits its attacker with its hands, scoring them with its claws. An inexperienced jaguar often retreats from such treatment. Anteaters are also hunted by man.
For most of the year, anteaters are solitary, seeking partners only in the breeding season. The female gives birth to a single young, weighing about 1.5 kg, after a gestation period of 190 days. She carries it on her back, and when the infant is hungry, it crawls round to the front to suckle, and then returns along the mother's tail to her back.

Apara or Brazilian Three-banded Armadillo

APARA or BRAZILIAN THREE-BANDED ARMADILLO
Tolypeutes tricinctus

The Apara or Brazilian Three-banded Armadillo reaches a length of only 30 cm. It dwells in bush-covered sites among large boulders, or in holes between roots of trees and bushes, and does not dig its own burrows. It is the only species of armadillo which can curl into a complete ball when in danger. The segments of its hard carapace and the bony plates on its head fit so tightly together that the animal forms a perfectly closed ball, hiding its abdomen and tail.
The Apara has an acute sense of smell, which helps it to hunt insects, worms, molluscs, or even small lizards, and to find birds' eggs after dusk. It also eats fallen berries. The female produces a single young after a gestation period of 5 — 6 months. The real length of gestation during which the embryo develops in her body, is only 2 — 2.5 months because after fertilization, implantation in the wall of the

uterus does not take place for many weeks. This phenomenon is called delayed implantation.
The Apara is preyed upon by large carnivores, especially by jaguars, whose powerful teeth can easily penetrate the carapace.

Nine-banded Armadillo

NINE-BANDED ARMADILLO
Dasypus novemcinctus

The Nine-banded Armadillo is distributed from the south of the United States over Central America to northern Argentina. Its carapace, which is composed of nine bands, is 40 cm long, and the tail measures 15 cm. As with other armadillos, the carapace is made of bony plates covered with a horny layer, and the bands are connected by soft flexible skin. The head is also covered with thick bony plates. The forelimbs are adapted for digging, and the digits have powerful claws.
The Nine-banded Armadillo lives in bushland where it digs burrows more than 1 m deep, and with corridors over 5 m long. The den has up to 12 exits through which the animal can easily escape. It leaves its burrow at night, briskly foraging for food. It searches for anthills and termitaria, breaking the walls and eating the inhabitants. It also feeds on other insects, worms, molluscs, birds' eggs, small birds and mammals, as well as eating fruit and green shoots.
For most of the year, armadillos are solitary, forming pairs only in the mating season. The gestation period extends over 4 months, but the young develop only for 2.5 months. As in several other mammals, the ovum is not at first implanted in the wall of the uterus, and it stops developing. This phenomenon is called delayed implantation. The female always gives birth to 4 young, always of the same sex, for they are produced from a single, fertilized egg which subsequently divides. The skin of the infants is very delicate. The carapace plates begin to harden after a few weeks. The offspring is suckled for a month.
The Nine-banded Armadillo is preyed upon chiefly by canine and feline carnivores, who turn the animal on its back to get to its soft abdomen. The Indians hunt it either at night with dogs, or dig it out of its underground burrows.

Mara or
Patagonian Cavy

MARA or PATAGONIAN CAVY
Dolichotis patagona

The Mara or Patagonian Cavy is an interesting South American rodent which closely resembles a hare. It is a substantial animal, measuring up to 50 cm. It has a short tail, only 5 cm long, short fur, long legs and hoof-like claws. It inhabits arid areas with coarse grass and semidesert regions of Patagonia, where it is bitterly cold in winter. Maras live in small groups of up to 40. They are diurnal and graze after sunrise, or browse on leaves and soft twigs. When in danger, they can run at a speed of 40 km per hour, making leaps 2 m long. When their movement is more leisurely, they take short, slow hops. As they are good to eat, they are often hunted. For this reason maras have almost disappeared from some localities and survive in remote places only on account of their skill and swiftness. Dogs cannot easily catch them in brushwood because they zigzag adroitly and prevent their pursuers from running at full speed. During the day, they bask in the sun, some members of the group always on the alert for danger. Maras breed twice a year. The female digs herself a burrow and after a gestation period of 58 days she bears 1 — 3 young. They are well-developed, with fur and open eyes, and they are active from birth. They stay with the female for 9 months before becoming totally independent, but feed on plants from the age of one month. Maras have a long lifespan of 15 years.

Aperea
or Brazilian Cavy

APEREA or BRAZILIAN CAVY
Cavia aperea

Few people know that the familiar guinea-pig came originally from South America. The wild species, Aperea, or the Brazilian Cavy, lives both in lowlands and mountains from Guyana to Argentina, sometimes at heights of up to 4 000 m. It occurs in several subspecies, considered by some zoologists to be independent species. This rodent reaches a length of 30 cm and a weight of 0.5 kg. It is a resident of open grasslands, plains, and slopes overgrown with bushes. It feeds on all kinds of vegetation, nibbling shoots, grass, bark, fallen fruit and seeds. It occasionally also eats molluscs and insects, and even meat from carcasses.

In some places, apereas live in large colonies in thickets or underground in burrows. The native Indians search out these colonies and catch the apereas by the hundred. In some mountain settlements they provide food the whole year through. Apereas are also preyed upon by birds of prey and carnivores but fortunately they breed very quickly. The female usually gives birth to 2 young several times a year. They are born after a gestation period of 62 — 70 days and can see immediately after birth. They soon begin to run, and after a week they begin to eat leaves though they suckle for another 2 weeks. Apereas are mature after 8 — 9 months and live to be 6 years old. Several forms of this rodent have been bred in captivity, including angora and long-haired varieties.

AZARA'S AGOUTI
Dasyprocta azarae

Azara's Agouti is another interesting rodent of the South American continent. It inhabits bushland and the edges of forests in southern Brazil, Bolivia and Paraguay. It is about 45 cm long, has a stunted tail a mere 1.5 cm long, and relatively long legs. It lives in pairs or in families, or sometimes in large groups. It is nocturnal, sheltering in hollow trees or among stones during the day, and setting out to forage after sunset. It seeks various kinds of fallen fruit, and its sharp incisors can perforate even the tough shells of coconuts. Agoutis sit erect to eat, holding the food in their forepaws. They occasionally take nestlings, and catch small iguanas, insects and molluscs.

After a gestation period lasting 104 — 106 days, the female usually produces a single young, though sometimes she has twins. At birth the young have open eyes, and fur, and they start to run the following day. Agoutis have a lifespan of 15 years. Their natural enemies include carnivores, raptors and snakes.

CHINCHILLA
Chinchilla laniger

The Chinchilla is well known the world over for the beauty of its fur. This rodent lives in barren rocky areas high in the Andes Mountains. It has the finest fur of all the mammals, dense, pearly grey, silky fur, which is much prized and is very valuable. As this small animal is only 40 — 45 cm long (including 12 cm for the tail), the fur of over 100 is needed for just one coat. For this reason the Chinchilla has been hunted to extinction in many places. Huge colonies of several thousands were once found in Chile, Peru and Bolivia, but they no longer exist. Even at the time of the Incas, chinchilla pelts were used to make special garments for the rulers and priests. From the end of the 18th century, pelts were exported in tens of thousands to Europe, and the number of chinchillas steadily declined. According to the latest reports, chinchillas in the wild now occur only in a small area in Bolivia. Nowadays they are bred on farms throughout the world, but since they are not very prolific, their fur is still very expensive.
Chinchillas live in pairs, and the male takes part in parental duties. The female usually gives birth to 2 young in April or May after a gestation period lasting about 110 days. The young are born with open eyes and start running about on the following day. They are suckled for 2 months and then feed on grass, shoots and roots. Chinchillas make burrows or live in rocky caves. They begin to forage after sunset and are adept at climbing the rocky slopes on which they live.

VISCACHA
Lagostomus maximus

The Viscacha is found in bushland ranging from southern Bolivia to central Argentina. This rodent is

Azara's Agouti

65 — 70 cm long, including the tail, which measures 17 — 19 cm. It is found mainly in pampas, at heights of about 700 m, but it may also be seen in mountains up to 5 000 m.
Viscachas live in large colonies of up to several thousands which occupy subterranean tunnels with many entrances. The pampas is dotted with the many holes dug by viscachas. These holes become overgrown so that they cannot be seen and are a danger to horses, which can easily break their legs in them. Colonies can be identified from a distance by the heaps of

Chinchilla

Viscacha

bones lying around the burrows. The bones are the remains of the skeletons of various dead animals. They are collected by the viscachas which gnaw them to file down and sharpen their powerful teeth. They often drag along cows' horns, stones or bits and pieces discarded by man. During the day they hide in their burrows, setting out after sunset to feed on grass, roots and leaves. A large colony can graze off almost all the grass in the neighbourhood, but large colonies have become rare, for viscachas are hunted for their fur, and have also been exterminated in some places as pests. They are not very prolific. The female produces 2 — 3 young once a year after a gestation period of 145 days. The young can see immediately after birth, and are already furred. They begin to walk after only a few hours, and are suckled for 6 — 7 weeks. Their average lifespan is 7 years unless the animal falls prey to predatory beasts or birds.

VICUÑA
Vicugna vicugna

Llamas are among the best-known of the South American animals. One of them, the Vicuña, is a wild species living in the mountains of southern Ecuador, Peru and Bolivia, at heights of 3 500 — 6 000 m. It has a very fine, long hair which protects it from the cold, for at night and in winter, the temperature falls way below zero. However, the quality of its wool has proved fatal for the llamas. The Vicuña has been hunted to extinction and is now found only in a few localities. It has become an endangered species. It stands 70 — 110 cm high at the shoulder and it is pale yellow-brown in colour. It is probably the ancestor of the domesticated form called the Alpaca.
Vicuñas frequent grassy slopes, where they feed almost entirely on grass. They are expert at climbing

Vicuña

Guanaco

the steep slopes and easily escape their natural enemies. They form small herds of 6 — 8, led by a strong, cautious male. The females with their young often form independent groups, and young males form groups of their own until they reach maturity. After a gestation period lasting about 300 days, a single young or twins are born, which can follow their mother the second day after birth. Young vicuñas are sexually mature after 2 years.

GUANACO
Lama guanicoe

The Guanaco is the most widespread of the wild llamas of South America. It is the ancestor of the Llama, a domesticated race used by the native Indians as beasts of burden. Llamas also provide meat, and their fur is used to make the blankets so necessary in the chilly Andean nights and cruel winters. The Guanaco is a well-built hoofed mammal related to the camels, which reaches a height of 90 — 130 cm at the shoulder, a length of 2.25 m and a weight of 60 — 75 kg. It is found in the Andes from northern Peru to Tierra del Fuego, on grassy steppes and semideserts up to a height of 4 000 m. Guanacos feed on grass, green shoots on stunted bushes, leaves and fruits. They live in small herds, each led by a robust male, which stands guard near the grazing animals. When he notices danger, he warns the others, and the herd rushes to safety, dexterously moving on the stony slopes. These animals are also able to swim well. They have an acute sense of hearing and keen eyesight, which helps them to survive in some localities. Guanacos have special skin glands on their hindlegs, which exude a secretion used to mark out territories. These territories are fiercely defended by the males, and in the rutting season rivals engage in savage duels. After a gestation period of 330 days, the female produces a single young, or sometimes twins. These are very active by the next day and are able to accompany their mother. They are mature after 3 years and live to be up to 25 years old.

COMMON RHEA or NANDU
Rhea americana

The Common Rhea or Nandu is distributed from north-eastern Brazil to central Argentina. It roams through the grassy pampas among bushes and isolated trees. It stands about 150 cm tall and weighs about 25 kg. It cannot fly but can run at a speed of 50 km per hour when in danger. The sexes are similarly coloured, but adult males have blacker feathers at the base of the neck. For most of the year, these birds wander through the countryside in small flocks, foraging for grass, leaves, shoots and fallen fruits. They also eat insects, molluscs and tiny vertebrates, swallowing the prey whole, and sometimes gulping down a few pebbles to crush the food in the stomach. In the courting period, groups of up to 30 males gather to hold 'tournaments' in front of the watching females. They usually fight just for display, and without hurting each other. They also run in circles, describing sharp turns and balancing with their wings, and they make low, bubbling calls. The male later builds a nest about 1 m across, in the grass. It is usually situated under a bush, and is sparsely lined with stalks. It is often filled with eggs, laid by several females. A clutch usually contains 12 — 30 yellowish eggs. The male incubates them for 42 days until they hatch, and then cares for the young chicks, shepherding them about and sheltering them under his wings at night. He also shades them during the midday heat. The young live on the egg yolk for a few days and then begin to catch insects and peck at green leaves. They become independent after 4 months, but gather into flocks with other families. They are fully mature after 2 years. Rheas are preyed upon by man, as well as by carnivores and raptors, and they no longer exist in some areas. They are hunted by gauchos on horseback, using a weapon called a bola. This consists of two or more heavy balls on the ends of a strong cord. When thrown it becomes tangled in the legs of the victim. At first the birds run away, changing direction to slow down the horses. Then suddenly, they lie down in the tall grass, so that they can no longer be seen. The South American Indians also collect and eat the eggs of these birds.

MAGELLAN or UPLAND GOOSE
Chloephaga picta

The Magellan or Upland Goose is common along the ocean coast of South America from Chile to Tierra del Fuego. It occasionally occurs on the Falkland Is-

Common Rhea or Nandu

lands, and it visits the deltas of the Rio Negro and Rio Colorado. It prefers grassy areas near the coast and seldom enters the water. It weighs 2.5 — 3.2 kg, and is characterized by distinct sexual dimorphism, the male being predominantly whitish, and the female brownish in colour.

The nesting season in Patagonia begins in November or December. Pairs take up occupation of their territories and defend them fiercely against intruders. The male chases away other species of birds, often bigger than himself. The nest is built by the female, usually under a bush, in tall grass or among boulders. It is lined with fine grass or leaves, and the clutch of 8 — 15 pale brown to slightly reddish eggs is surrounded by down. The female incubates the eggs for 30 — 32 days, while the male guards the nest. The blackish-brown, white-spotted goslings are cared for by both parents. After nesting, the pairs merge into flocks and migrate to warmer northern regions, only occasionally hibernating in their homeland. They feed on plants, seeds and berries.

Magellan or Upland Goose

ANDEAN GOOSE
Chloephaga melanoptera

The Andean Goose occurs in the Andes mountains of South America, from central Peru southwards to Patagonia, at a height of 2 000 — 5 000 m. It is found on steppes and grassy slopes, usually near small lakes, beside mountain springs at the foot of glaciers, and in valleys with running streams. Both sexes are identically coloured. The male weighs up to 3.6 kg, but the female weighs only 2.8 kg.

The Andean Goose lives in pairs, each pair having large nesting territory from which the male drives every intruder. The female builds the nest in a rocky cleft or under dense, low bushes, and lines it carefully with down. She incubates her clutch of 6 — 10 cream-coloured eggs for 30 days while her mate guards her and chases away intruders. He attacks by hitting them with his wings, beak and claws. The chicks are black, brown and white spotted. As soon as they have dried off a little, the parents take them into lagoons among the swamps, for safety. At night and in bad weather, the female covers them with her wings and the male stays on guard. When the dry, cold season begins, the pairs gather in flocks and migrate northwards.

Andean Goose

Turkey Vulture

The Indians catch these young geese and keep them as domestic poultry. The Andean Goose makes a chattering call, and feeds on grass, shoots, seeds and berries.

TURKEY VULTURE
Cathartes aura

The Turkey Vulture is one of the most common American birds of prey. Its homeland is the open countryside stretching from southern Canada across the United States and Central America to the southern tip of South America. It also lives on the Falkland Islands, and is most widespread in the tropics. It is 75 cm long and has a wingspan of about 180 cm. Both sexes are alike in coloration. This striking bird frequents rocky localities, deserts and fields. In autumn, it migrates from the north southwards, often in flocks of several hundreds. During the year, it lives in small groups and visits towns to feed on garbage, particularly scraps of meat. Scores of these vultures often circle for hours above settlements.

Fresh carcasses form the bulk of their food. They often fly to the coast to feed on the remains of fishermen's catches, and sometimes they hunt small vertebrates, such as snakes, frogs and rodents. They also eat eggs and fruit.

At nesting time the Turkey Vulture looks for a small cave, preferably with two exits, in a high rocky locality. It may also nest in a large hole in a tree or in a nest deserted by other raptors. It never builds its own nest but lines the nesting cup with scraps of hide from carcasses. The female lays 2 — 3 whitish eggs with brown spots of various shades. The clutch is situated in the darkest corner of the hole or cave, and is incubated by both parents. The young hatch after 38 — 41 days and they are fed on partially digested food from the parents' crops. Young vultures open their eyes on the first day, and after 4 weeks they bask at the edge of the nest, usually in the morning or in late afternoon. They leave the nest after 70 — 80 days, but if food is in short supply, they take longer to develop. Several pairs often nest near each other, in a colony. The nesting season in the tropics begins in February, in Florida it begins in March, and in the north, in June.

This raptor has a hissing, hoarse voice, and makes grunting noises in the nesting season. When it is asleep, its temperature drops to 34° C, so its body loses less heat energy. When it wakes up, the temperature rises to its usual 39° C.

BLACK VULTURE
Coragyps atratus

The Black Vulture is distributed from central Patagonia across South and Central America to the United States. It is common in Ohio and Arizona, and occurs in the state of Washington. In autumn, it leaves the northern regions for South America, and in November, flocks of several hundreds cross the Panama Canal. This raptor is about 65 cm long and weighs 2 kg. It is one of the best known birds of the tropics. It visits cities but is found even in deserts. These vultures circle the sky in large flocks, looking for food. Up to 300 birds may gather round a large carcass, some describing circles in the air, others slowly walking around the carcass, and the rest tearing off hunks of flesh. When replete, the vultures take long rests in the trees, basking in the sun with half-open wings. The huge flocks roost on high rocks, or in trees. If no carrion is available, they hunt small animals along the coast, especially young sea birds, or wait for hatching sea turtles. Young turtles have soft carapaces and are easy prey. Black Vultures also favour soft fruit and oil palm fruits.

They remain in colonies even in the breeding season, the pairs building nests close to each other, preferably in caves. They often nest on cliffs or in holes low down in trees. The female lays 2 eggs, mostly grey-green in colour, with several large brown spots. Both partners incubate the clutch for 32 — 39 days. The female lays her eggs in a dark corner, but later pushes them towards the entrance of the cave. The chicks are fed on regurgitated food — up to ten times a day to begin with, but only once or twice after about 40 days. After 75 days, the young vultures begin to fly. On the equator, nesting takes place from February to June, while in Argentina it is from October to November, and in Trinidad, from November to January. In the United States vultures nest from March to April.

Black Vulture

ANDEAN CONDOR
Vultur gryphus

The Andean Condor is the largest raptor in the world. The male weighs as much as 12 kg and is 1 m tall, while the female weighs 10 kg. It has a wingspan of up to 3.2 m. This condor lives in the Andes, from western Venezuela to Argentina and is also found on coastal cliffs in Patagonia. It is especially common in Colombia, where flocks of up to 60 can be seen. The Andean Condor lives in pairs or solitarily. Among the high mountains, the birds circle for hours in the sky, searching for prey. They look in particular for large game lost over precipices. Often they feed on dead llamas, and they also hunt for small animals or take the eggs and nestlings of sea birds. Although adventure books frequently suggest that the condor carries off large prey in its claws, this is not the case. Nor can it carry a child. Its claws are very powerful, but they are straight and blunt, so the bird cannot grab its prey.

Condors make their nests high up on rocky platforms, several pairs often nesting together. The nesting cup is lined with scraps of hide. The female lays a single egg, which is incubated by each partner in turn for 54 — 58 days. Both parents care for the chick, at first feeding it on regurgitated food. The young condor leaves the nest after 6 months, and stays with its parents for another 6 — 8 months. A pair nests only once in two years. Young condors are brown all over and lack the characteristic protuberances above the beak. They assume their adult coloration when they are 4 — 5 years old.

WHITE-TAILED KITE
Elanus leucurus

The White-tailed Kite is distributed in the southern part of the United States and in South America as far south as southern Chile and central Argentina. It is particularly common in the tropics. It is about 35 cm long and has a wingspan of 100 cm. It flies above savannahs and desert areas, looking for prey from a height of 15 — 20 m. Sometimes it seems to stop in mid-air before suddenly swooping down. It

Andean Condor

feeds mainly on small rodents, but also catches reptiles and amphibians, and it plummets down low above the grass to flush out insects and catch them swiftly in its claws.

Kites usually live in pairs, though they may be solitary outside the nesting season. During the day scores of them often gather together, up to 50 perching in a single tree.

In the courtship period, the male flies high above the savannah, circling and turning in the air, and at the same time making whistling sounds. The pairs defend their territories, which are several hundred metres apart. The nest is built by the female, but the male accompanies her, occasionally bringing building material, such as dry twigs, broken off with his beak. The nest is usually situated about 25 m above the ground, and the nesting cup is lined with dry grass. Throughout the nesting period, the male becomes responsible for catching food for his mate. When he brings it, she leaves the nest and meets him about 20 m above the ground. Both birds make loud cries while their bodies turn vertically and the female takes the food.

The female lays 4 — 5 off-white eggs, densely covered with brown spots, and incubates them for 29 days while the male perches nearby. She divides the prey into portions for the chicks until they are about 25 days old and are able to tear it up for themselves. The chicks leave the nest after 35 — 40 days, but return at night-time for a little longer. The parents continue to feed them for another three weeks while teaching them to fly. After this they usually build a new nest and rear a second brood.

WHITE-TAILED HAWK or WHITE-BREASTED BUZZARD
Buteo albicaudatus

The White-tailed, or White-breasted Buzzard, is widespread throughout tropical South and Central America and in Texas. In winter, it sometimes travels south, but it is not migratory. It is up to 55 cm long and has a wingspan of 120 cm. The sexes are similar in coloration. This hawk mostly inhabits open country where there are solitary trees or rocks. It perches

on these high vantage points from which it takes off to pursue its prey. It feeds on rodents, snakes, lizards, cicadas and beetles, and occasionally snatches flesh from the carcasses of large game.
In the courtship season, the male delivers a rapid whistling note as he circles in the air and executes sharp turns. The nest, which is about 1 m across, is constructed in a tree at a height of only 1.5 — 3 m above the ground. It is made from fresh twigs and the nesting cup is lined with moss or dry grass. The female lays 2 — 4 eggs, usually pure white, though sometimes with small brown spots. Both partners incubate the clutch for 28 — 31 days, the female spending most time on the nest. The newly hatched young are fed solely by the female, on food brought by the male. Later he also feeds the chicks. After 45 days, the young hawks are capable of flight and leave the nest, but their parents continue to feed them for another month.

White-tailed Kite

GALÁPAGOS HAWK
Buteo galapagoensis

The Galápagos Hawk lives in the Galápagos Islands. It no longer survives on San Cristobal and Floreana, but it is quite common on the islands of Santa Fé, Fernandina and Isabella. It is protected, but the entire population of this exclusively Galápagos raptor is estimated at only two hundred. This hawk feeds mainly on dead fish found along the coast, and on insects and young iguanas. It occasionally hunts snakes and rats, which have been introduced to the islands, and it catches red crabs, left behind on the rocky shores as the tide ebbs. It also feeds on birds' eggs.
The nest of this raptor is situated in a low tree, or on the ground on a lava rock. It is made from mangrove branches, and is lined with grass, leaves and pieces of bark. The female lays up to 3 greenish eggs, and both partners incubate them for 37 days. The newly hatched young are at first fed by the female, while the male hunts for prey, but older chicks are fed by both parents. Only a single chick usually survives from each clutch and it leaves the nest after 50 — 60 days. The nesting season in the Galápagos Islands can be at any time of the year.

*White-tailed Hawk
or White-breasted Buzzard*

Galápagos Hawk

CRESTED CARACARA or MEXICAN EAGLE
Polyborus plancus

The Crested Caracara or Mexican Eagle can attain a length of 55 cm and a wingspan of 120 cm. Both sexes are identically coloured. This raptor is distributed from central Florida over Central and South America to the Falkland Islands. It lives in open country or at the edges of forests, in lowlands and in mountains up to 3 000 m. It has long legs adapted for walking, and it seeks food on the ground, hunting small vertebrates and insects, and collecting soft fruits. The area around the nesting territory is fiercely defended by these birds, who chase even large birds of prey from their nests and take both eggs and nestlings to eat.

In the courtship period, the male circles high in the air. The birds often greet each other with loud, croaking sounds, bending their heads backwards so far that they touch their backs with the nape of their necks. The nest is made of branches, and is lined with turf, grass, pieces of hide, and sometimes even tiny bones. The female lays 2 — 4 pinkish-white eggs, sometimes with black or brown spots. Both partners take turns to sit on the clutch for 28 days, and both look after the brood until at the age of 2 months, the young leave the nest.

The natives dislike these birds, for they often take domestic poultry. Although the Indians pursue them, they rarely manage to kill them, for their weapons are primitive.

CRESTED SERIEMA
Cariama cristata

The Crested Seriema is a sturdy, long-legged bird, about 95 cm long and 85 cm tall. It inhabits vast bush-covered savannahs with solitary trees, ranging from central Brazil to northern Argentina and east-

Crested Caracara or Mexican Eagle

ern Bolivia. This bird is an indifferent flier and moves mainly on the ground. It is able to run very quickly, but it prefers to seek refuge in a tree if it is chased by a dog or a wild animal. At night it roosts on the branches of trees, to avoid terrestrial predators. The Crested Seriema lives in pairs or in small family groups. During the day, it wanders throughout its territory, skilfully hunting lizards, amphibians and small mammals, or collecting seeds and berries, and pecking at shoots. When catching large prey like snakes or lizards, it grasps the creature in its beak and kills it by smashing it against the ground or on a stone. Then it swallows the prey whole.
The voice of the Crested Seriema is piercing, and carries over a long distance. It uses it throughout the year and not just during the nesting season when the males engage in duels and the winner takes a mate and founds a family in his territory. The pair builds a nest of dry twigs in a low tree or bush, 2 — 3 m above the ground, and the female lays 2 pale brown eggs with dark brown, black and purple spots. Incubation takes 25 — 26 days, and is carried out mainly by the female. Both parents care for the young, which leave the nest when they are 14 days old and follow their parents down to the ground. At night they hop up on to low branches to roost. They become independent at the age of 3 months, but remain with their parents until the next nesting season, or form a small group with other young birds.

Crested Seriema

Red-masked Parakeet

RED-MASKED PARAKEET
Aratinga erythrogenys

The Red-masked Parakeet is about 33 cm in length, and lives in western Ecuador and western Peru. Young birds lack the red head coloration. This parakeet is resident in dry mountain localities, most frequently at a height of 2 300 m. It particularly favours slopes with bamboo thickets. Scores of these parakeets often sit on a single bamboo stalk, their weight bending it almost to the ground. For most of the year, these birds travel together in flocks, sometimes of several hundreds, and their cries can be heard far afield. In the nesting season the Red-masked Parakeet forms pairs, which often gather in colonies of 50 — 60 birds. They build nests in clefts or crevices in vertical rocky walls. The female lays 4 — 6 eggs and incubates them for 25 days, while the

Thick-billed Parrot

male brings her food. Both parents take care of the young, which leave the nest after 8 weeks. The families then merge together and roam throughout the countryside. This parakeet feeds on shoots, seeds and juicy fruits.

THICK-BILLED PARROT
Rhynchopsitta pachyrhyncha

The Thick-billed Parrot is instantly recognized by its stout bill. It body is 40 cm long and both sexes have the same coloration. This heavily built parrot inhabits north-western and central Mexico, and occasionally appears in Arizona. It lives on plateaus in sparse pine forests, where it feeds mainly on the seeds of tough pine cones, easily peeling off their scales with its thick bill. It can also crack nuts, but it looks for various soft fruits as well. It comes down to the ground when the forest berries are ripening, but it is always cautious, picking up a berry and immediately soaring back to a safe treetop.

The nest is built in a hollow in a pine tree, usually over 10 m above ground. The nesting cup is not lined. The female lays 2 — 4 eggs and incubates them for 25 days. The male feeds her several times a day, regurgitating seeds for her. The young are fed by both parents and leave the nest when they are 50 — 58 days old. They are fully coloured by this time, but are more subdued in appearance than the adults. The families roam throughout the countryside, and in winter they fly to lower-lying regions and valleys at heights from 1 550 — 1 700 m.

BURROWING PARROT
Cyanoliseus patagonus

The Burrowing Parrot is distributed in Chile, Uruguay and Argentina. It is about 43 cm long and the sexes are alike in coloration. In winter, these parrots fly from mountainous regions to protected sites in the valleys. They form large flocks of up to several hundred birds and fly in areas north of Buenos Aires, occasionally reaching Venezuela. These parrots can withstand frost for a period of several weeks.
In the breeding season, the Burrowing Parrot forms colonies. The pairs dig out burrows in clay or sandy cliffs and in river banks. The entrances are close together, and each corridor is about 120 cm long, terminating in a nesting cup. The parrots sometimes build nests in a rocky cleft or low in a tree trunk. The female lays 2 — 5 eggs and sits on the clutch for 24 — 26 days, while the male feeds her on seeds, fruits and shoots, and occasionally on insects. Large flocks of these parrots sometimes raid corn fields, causing considerable damage to the crops. The young parrots are fed by both their parents. They leave the nest after 8 weeks, fully feathered and fully coloured, differing from the adults only in having a whitish upper part to the beak, instead of black.

SAPPHIRE-RUMPED PARROTLET
Touit purpurata

The Sapphire-rumped Parrotlet is a small species, reaching a length of only 17 cm. The female has less conspicuous hues on her wings and tail. This parrot is common in wooded savannahs and in sandy localities along the edges of forests, ranging from south-eastern Colombia to the Amazon. For most of the year, it wanders in small groups in the treetops, looking for food. It feeds on soft fruits, seeds and shoots, and is particularly fond of mangoes. The nest is usually built in a hollow tree. Sometimes, however, the birds dig a small cavity in a prominently situated termitarium. Here the female lays 3 — 5 eggs, and incubates them for 25 days, while the male feeds her. The chicks are cared for by both parents, and leave the nest after 8 weeks.

Burrowing Parrot

Sapphire-rumped Parrotlet

Red-billed Parrot

RED-BILLED PARROT

Pionus sordidus

The Red-billed Parrot exists in six subspecies. It reaches a length of about 28 cm, and both sexes are alike in coloration. Its range of distribution is extensive, covering Colombia, Venezuela, northern Bolivia and eastern Peru. It lives mainly in mountain forests at a height of about 2 000 m. Outside the nesting season, it travels in flocks of about 40. It usually flies to lower-lying localities at about 300 m to forage, feeding predominantly on seeds, fruits and shoots. It crushes small nuts with its thick bill, and picks out the kernels. In the nesting season, the pairs take up occupation of a territory and look for a tree with a suitable hole in it. The female lays 2 — 5 eggs and sits on them for 27 days. The young leave the nest when they are 2 months old. In some areas, this parrot is quite abundant, but it is quiet and rarely heard.

TROPICAL SCREECH OWL

Otus choliba

The Tropical Screech Owl measures 19 — 23 cm and is distributed from Costa Rica to Buenos Aires in Argentina. It is a resident of dry savannahs dotted with bushes and occasional trees, and of dry wood-

Tropical Screech Owl

land. It is solitary for most of the year, roaming throughout its hunting territory. It chooses a permanent site in a tall treetop and sets out to hunt at dusk. Its prey consists mainly of small vertebrates and insects, but it sometimes catches prey larger than itself. It forms pairs in the nesting season, but does not build its own nest. Instead, it acquires a nest deserted by some other bird in a suitable treetop, or takes up residence in a convenient hole. The female, coloured like the male, lays 2 — 5 eggs, which she incubates for 24 days. Her mate brings food for her to a spot near the nest. The young are cared for by both parents. After 4 weeks, the chicks leave the nest to perch on branches where they continue to be fed. Only after a further 2 weeks do they venture to take their first flight.

FERRUGINOUS PYGMY OWL
Glaucidium brasilianum

The Ferruginous Pygmy Owl is a small owl, about 16 cm long, and having a wingspan of 48 cm. Both sexes are identically coloured. Two colour forms exist in the wild, one of which is predominantly grey, and the other reddish. This owl is distributed from south-eastern regions of the United States across Mexico and Central America to southern Argentina. It also lives on the island of Trinidad. It is most common in Brazil, though there its distribution is less continuous. It lives in open semidesert localities and on wooded river banks in lowland areas and on hills up to a height of 1 200 m. In Mexico, the pygmy owl often perches on tall cacti, basking in the sun, although it spends most of the day among dense branches or in holes in trees or cacti. It occasionally catches young iguanas, especially anoline lizards, and sometimes it attacks small birds. It forms pairs in the nesting season. The female lays about 4 eggs in a hole and sits on them for 26 days. The young are fed on insects by their parents, until they leave the nest at the age of one month.

Ferruginous Pygmy Owl

ELF OWL
Micrathene whitneyi

The Elf Owl is a mere 13 cm long, and has a wingspan of only 37 cm. Both sexes are alike in coloration. It is a native of cactus-covered desert and semidesert areas of central Mexico northwards to the south-eastern part of the United States. It can also be found in thinly forested areas up to a height of 2 300 m, and is quite abundant in some places. During the day, it rests in the hollow of a giant cactus, setting out to hunt after sunset. It forms pairs in the nesting season. The female lays 4 — 6 white eggs in a hole in a cactus, and incubates them for 27 days, while the male brings her food. The young are fed by both parents, and leave the nest when they are 4 weeks old.

Elf Owl

SMOOTH-BILLED ANI
Crotophaga ani

The Smooth-billed Ani is abundant in bush-covered localities from southern Florida across Central America and the West Indies to Argentina. It is about 33 cm long. Outside the nesting season, flocks of these birds visit city parks and gardens. In flight, they often deliver a jingling, rather pleasant call. They live near the coast or in damp grasslands frequented by herons. The herons disturb insects when walking in the grass, enabling the anis to catch them. Anis also feed on other invertebrates, seeds and berries.

The breeding season varies from place to place. The nest is built in a dense bush or tree, up to 7 m above the ground. It is a large, bowl-shaped structure, made of grass, and the nesting cup is lined with tiny green leaves. The nest is usually built by only one female, although several birds have been seen to take part in the construction of a large nest. Normally three females lay their eggs in the nest, the clutch totalling 10 — 15 blue-green eggs. It is incubated by just one of the females for 13 days. The young leave the nest after 15 — 17 days and stay with the flock. Older independent chicks from the first nesting sometimes feed the younger fledglings, and the whole group participates in rearing the brood.

Smooth-billed Ani

GREAT POTOO
Nyctibius grandis

The Great Potoo is about 50 cm long, and is a relative of the nightjar. It is found in savannahs, on the edges of forests, and in plantations, from Panama to Peru and Bolivia. This is a solitary bird. During the day, it perches on a branch in a vertical position with its head up and its tail pointing down. In this way it blends with the background, perfectly imitating an old dry stump. After sunset, it makes forays from

Great Potoo

a branch, flying silently and skilfully just above the ground and catches nocturnal moths and beetles on the wing, before returning to its perch. In the breeding season, it sometimes makes a drawn-out call and sometimes just a trilling phrase. At this time it forms pairs. The female lays a single egg at the bottom of a hole in a tree, seldom more than 1 m above the ground. The egg is white with several purple and brown spots. Both partners share in incubation for a month, and if an animal approaches, the nesting bird stands up menacingly, with opened beak, to threaten the intruder. The chick sits rigidly in its nest and is probably fed only by the female, who brings it insects. After some 40 days, the young potoo leaves the nest to learn to fly and hunt.

Bee Hummingbird

BEE HUMMINGBIRD
Calypte helenae

The Bee Hummingbird is the smallest of the hummingbirds and this makes it the smallest species of bird in the world. It is only 6.5 cm long and weighs a mere 2 g. The female is green above and whitish below, and has white spots on her tail feathers. This hummingbird inhabits open bush-covered localities and the fringes of forests in Cuba and the adjacent Isla de Pinos. It also visits city parks and gardens, attracted by the flowers, and may fly through an open window, if there are flowers inside. It will dart in, stop in the air, suck nectar for a few seconds, and vanish, emitting a wailing cry. This species also catches tiny insects, especially flies. In the nesting season, the female builds a tiny nest on a branch. Here she lays 2 eggs and incubates them for 14 days. The nesting habits of the Bee Hummingbird are similar to those of other species of hummingbird.

QUETZAL
Pharomachrus mocinno

The Quetzal is one of the best-known birds of Central America. It was once the sacred bird of the ancient Aztecs, and to them its beautiful plumage was more valuable than gold. This species dwells at a height of 2 000 — 3 000 m in humid tropical forests from Panama and Costa Rica to Mexico. It is about 36 cm long, but the courting plumage of the male is adorned with tail feathers over 60 cm long.
During the courting season, the male flies round his mate with his long ornamental feathers extended behind him like a train. Sometimes several pairs take part in nuptial displays in the same place. The male makes a fine whistling call which intensifies into a wailing cry. Each pair builds a nest in a dead tree

trunk, sometimes choosing a hole deserted by woodpeckers, and sometimes digging out a new hole in soft decaying wood. The hollow can be as deep as 30 cm, and is usually 5 — 20 m above the ground. Both partners help to build the nest and when it is ready the female lines the cup with fine slivers of wood, and lays 2 — 4 spherical, green-blue eggs. Both parents incubate the eggs, though the female spends most time on the nest. When the male sits on the clutch, he turns round and faces outwards, so that his long tail feathers are bent forward across his back and head, and project some 30 cm out of the hole. The young hatch after 17 — 18 days, and both parents feed them for a month on the nest, and for a further 14 days when they have left it.

Quetzals feed on fruits, insects, molluscs, spiders, and occasionally on small geckos. They often peck berries while in flight, vigorously fluttering their wings to support themselves in the air. They often rear two broods in a season, using the same nest. Outside the nesting season, they form small groups or fly singly. Each bird has its own particular perch on a dry branch, from which it takes off on its forays in search of food.

Quetzal

RUFOUS OVENBIRD or BAKER
Furnarius rufus

The Rufous Ovenbird or Baker is a typical bird of the pampas of southern Brazil, Argentina, Paraguay and Bolivia. It is about 20 cm long, and both sexes are identically coloured. For most of the year, ovenbirds live in small groups or in pairs. They frequent grasslands covered with bushes and occasional trees, and are plentiful near human settlements where cattle and horses are bred. Such places abound in insects, which form the bulk of their food, though they also eat spiders and molluscs. In the rainy season, when there is plenty of mud, each pair builds a nest made of earth, out in the open on top of a thick horizontal branch, on a wooden fence or building, or on rocks. Both partners bring balls of clay to the selected spot, where they crush and trample them with their feet. The birds build circular layers, strengthened with pieces of grass or straw, gradually constructing a symmetrical dome 20 cm tall, and 30 cm across. The walls are up to 4 cm thick, and at the bottom there is an entrance 8 cm high and

Rufous Ovenbird or Baker

Peruvian or Andean Cock of the Rock

5 cm wide, separated from the nesting chamber by an antechamber. The whole construction weighs 5 — 10 kg. Ovenbirds are expert builders and are said to make their sophisticated nest in 10 — 16 hours! As the clay dries, it hardens and forms a solid shelter for 2 — 4 white eggs, which are incubated by both parents for 18 days. The young are fed by their parents for 2 weeks in the nest and for 10 days after leaving it. The ovenbird often builds several nests which are then used by other species of birds.

Black-tailed Tityra

PERUVIAN or ANDEAN COCK OF THE ROCK
Rupicola peruviana

The Peruvian or Andean Cock of the Rock has a range of distribution stretching from north-western Venezuela and the Colombian Andes to Ecuador, Peru and northern Bolivia. This striking bird reaches a length of 28 — 32 cm. The female has a short crest and is orange-brown to dark brown in colour. An inhabitant of rocky walls above streams, and bush-covered slopes, up to a height of 1 300 — 2 500 m, it visits nearby mountain forests to seek berries and other fruits. It occasionally also catches insects. In the courtship period, the males present their nuptial dances in an open site, usually on a rocky platform or on patches of stones among the bushes. They hop, twist and display their plumage, while the females gather nearby, perching on branches or stones to watch. The pairs then look for a suitable place to nest. Several pairs may settle in the same rocky area and build their nests close together. The nest is a shallow structure made of clay, which is strengthened with twigs and lined with leaves. It is made by the female, who lays 2 white, black-dotted eggs, and incubates them for 3 weeks. Her mate helps her to feed and rear the brood.

BLACK-TAILED TITYRA
Tityra cayana

The Black-tailed Tityra is a cotinga. This bird is about 21 cm long, and is common from Venezuela and Guyana to southern Brazil, northern Argentina and Bolivia. It lives in open, bush-covered places with isolated trees, up to a height of about 1 300 m. The female differs from the male in having a dark green, black-striped back and a striped breast. For most of the year, tityras stay in small flocks and fly among the bushes, seeking berries and other fruits. They also catch insects crawling on leaves or on the wing. When resting, these birds perch on the tops of dead trees, and sleep in cavities in them. They form pairs in the breeding season, and build a nest in a hole in a tree, 12 — 35 m above the ground. The female sits on her 2 yellowish-white eggs for 21 days, while the male stays near her and warns her of danger, later helping her to feed the nestlings. Young birds are capable of flight when they are 25 days old, at which time they leave the nest. The parents usually also rear a second brood.

FORK-TAILED FLYCATCHER
Muscivora tyrannus

The male Fork-tailed Flycatcher has a distinctly elongated tail, 22 — 23 cm long, while its body measures only 8 — 10 cm. The female has a shorter tail of 17 — 20 cm. The area of distribution of this bird stretches from southern Mexico through Venezuela and Colombia, and across north-western Brazil to Bolivia and northern Argentina. Outside the nesting season, the birds from northern regions migrate southwards, while those from southernmost localities travel to the north. The Fork-tailed Flycatcher lives on flying insects, catching them adroitly in flight. In the nesting season, it forms pairs and takes up residence in savannahs, semideserts and deserts. It is very abundant, flying in parks and along the edges of thin forests, and pairs may take up residence near human habitations. The birds fiercely defend their territories, and the male chases away birds of prey. The nest is built by the female. It is a bowl-shaped structure of stalks, moss and leaves, situated in a tree or bush. The female lays a clutch of 3 — 4 white eggs, densely dotted on the thick end, and sits on them for 14 days. The young are fed by both parents, and young flycatchers leave their nest at the age of 16 days, though their parents continue to feed them for another 10 days. Two broods are usually reared each season.

Fork-tailed Flycatcher

RED-LEGGED THRUSH
Mimocichla plumbea

The Red-legged Thrush has its homeland in Cuba, the Bahamas, Hispaniola, Puerto Rico, Dominica and the small adjacent islands. There are several subspecies of this thrush which differ in coloration. This species is 25 — 28 cm long, and the female is distinguished from the male by her duller coloration. It inhabits both lowlands and hills, where it favours areas of brushwood in thin woodland, parks and gardens. It seeks food both on the ground and in trees, feeding mainly on insects and other invertebrates, though after the nesting period, it also eats berries. The nest is made in a bush or low tree and comprises roots, grass stalks, hairs and fibrous material. The female lays 3 — 5 eggs, usually green-blue, but sometimes with dark dots. The clutch is incubated for 13 days by the female, but both parents bring insects and larvae to their offspring. At the age of 20

Red-legged Thrush

Mockingbird

days, the young thrushes leave the nest and after a few days they begin to fly.

MOCKINGBIRD
Mimus polyglottos

The Mockingbird lives in West Indies, Mexico and in the south of the United States, nesting in California, Ohio, Iowa and Wyoming. It is a conspicuous bird, about 25 cm long and the female resembles the male, except that her colouring is less bright. In the nesting season, the pairs defend their territories ferociously, often chasing away birds stronger than themselves, and even attacking household cats and dogs. The nest is built by both partners from stalks, leaves, moss, wool and pieces of paper. The structure is situated in a bush or in a low tree with dense branches, usually at a height of 1 — 2 m above the ground, and takes 5 — 7 days to build. In the nesting season, the male often perches on an elevated spot, loudly singing on clear, moonlight nights. His attractive call is composed of melodious phrases, usually repeated three times. The female lays 3 — 6 greenish eggs with reddish-brown spots of various sizes. She usually incubates the clutch alone for 12 — 14 days, while the male brings her food, and sometimes briefly relieves her. The young are fed on insects, berries and pieces of soft fruit. The adult birds feed on seeds. After 10 days, the chicks jump out of the nest and shelter for several days among stones and tufts of grass. The parents feed them until the young begin to fly and fend for themselves. Some pairs rear three broods in a year.

CUBAN GRASSQUIT
Tiaris canora

The Cuban Grassquit inhabits open grasslands of Cuba, where it feeds on grass seeds, green plants, tiny insects and insect larvae. It is about 9.5 cm long. Young birds resemble the female, who has a greyish-brown nape. The nest is situated in dense branches and is made of grass, moss and small leaves. It is a spherical construction with an elongated, tubular, downward-facing entrance. It takes 4 — 6 days for a pair to finish their home. Sometimes, they do not finish the nest but begin to build again in a new situation. The female lays 3 — 4 white, or sometimes blue-green eggs, speckled with blackish-brown spots. She incubates the clutch for 12 — 13 days, while the male sits near the nest, driving away intruders. The young are fed for 13 — 16 days on the nest, before they begin to fly, after which the male continues to bring them food for a few days longer. By the age of 24 days, these birds are entirely independent.

MAGPIE JAY
Calocitta formosa

The home of the Magpie Jay is the area from south-western Mexico to north-western Costa Rica. Both sexes are alike in coloration, and measure 50 — 70 cm in length, of which half is occupied by the tail. This species lives in vast savannahs with scattered trees and bushes, or at the edges of sparse forests up to a height of 1 000 m. Outside the nesting season, it wanders in the neighbourhood of its home in small flocks, looking for food on the ground or in trees and bushes. It collects fruits, pecks shoots, hunts insects and spiders, and occasionally catches small reptiles. It has a piercing call, like that of a parrot. It forms pairs in the breeding season, and constructs nests of plant material on thorny bushes. The female lays 3 — 6 brown and grey-spotted eggs and incubates them for 18 days, while the male feeds her, and various other males also bring her food. These are probably single males showing instinctive nesting behaviour. The young leave the nest after 3 weeks and begin to fly after another few days, while still being fed by their parents.

Magpie Jay

Cuban Grassquit

GALÁPAGOS GIANT TORTOISE
Testudo elephantopus

The Galápagos Giant Tortoise is one of the best-known, largest and nowadays rarest tortoises in the world. It is confined to the Galápagos Islands, 1 000 km off the coast of Ecuador. When these islands were discovered in 1535, tortoises weighing as much as 250 kg were found in huge numbers throughout the islands. Each island had its specific subspecies, which implies that the animals had evolved in isolated communities over a long period. On the island where the tortoises predominantly

Galápagos Giant Tortoise

grazed, they developed normal carapaces, while on other islands, where they browsed on the leaves of bushes, their carapaces bent slightly upwards.
In the seventeenth century, the Galápagos Islands became the refuge of many ships, and their crews killed the tortoises for meat. Later, when ships stopped at the islands regularly, sailors caught the tortoises alive and kept them below decks, where they survived for months without food or water, and provided fresh meat throughout the voyages. The plunder continued for decades. The 1850s were marked by an upsurge in whaling. The whalers spent long months at sea, and when their food became scarce, they filled their stores with Galápagos tortoises. In an old whaling log Dr H.Townsend discovered that in the years from 1831 to 1869, during 189 visits to the Galápagos Islands, the whalers caught over 10 000 giant tortoises. The settlers who later reached the islands also fed on tortoise meat, and pressed the eggs to obtain oil. They brought dogs, cats, sheep, cattle and horses, and on many islands, the domestic animals quickly destroyed the vegetation on which the tortoises lived. Many of the dogs turned wild killed young tortoises. In 1825, a volcano erupted on the island of Narborough, and most of the tortoises there perished in the hot lava. Still the massacre continued. When in 1903 researchers from the Californian Academy of Sciences visited the islands, they found large numbers of tortoises only on three islands. As a result steps were taken to protect all the species on the islands, and keepers are employed by the government of Ecuador to this end. The remaining tortoises will hopefully be saved, although most subspecies are now extinct. Some 2 000 tortoises live on the island of Santa Cruz, where, to assure their survival, all the domestic animals have been deported. Several hundred more are found on Isabela Island.
On the larger islands, there are freshwater springs in the mountains inland. The tortoises from low-lying dry localities sometimes make trips lasting several days to reach the water. Every generation has followed the same paths which became so large that they were used by sailors as well. On arrival the tortoises gulp down the water and bathe, and after 3 — 4 days, they return to their grazing grounds. On some islands, the tortoises have to make do with pools of rainwater and juicy cacti.
In October, the female digs out holes and in them she lays round, white, hard-shelled eggs. Then she covers the eggs with soil which she beats down firmly. The young hatch after approximately 240 days, and crawl out to hide in the grass or bushes. Their only enemy is the Galápagos Hawk. The adult tortoises have no enemies except man, and when left alone can live to be 200 years old.

GALÁPAGOS LAND IGUANA
Conolophus subcristatus

The Galápagos Land Iguana inhabits some of the Galápagos Islands. It lives only on the land, and is a trusting animal, allowing man to approach it. Over the years this has led to its destruction, for its meat has always been popular with visiting sailors. Originally, these iguanas were abundant on the islands of Isabela and Santa Cruz. Since then, they have become almost extinct, and survive only on a few small adjacent islands which are seldom visited by man.

They are currently under protection like other rare species of the Galápagos Islands.
The Land Iguana reaches a length of 1.5 m. It has a yellowish head, a rust-coloured back, and dark brown flanks, but some islands are the home of greenish, grey-spotted specimens. All feature a low crest along their backs. Iguanas dig shallow burrows in crushed lava or volcanic sediments. They forage during the day, never moving far afield. They feed mainly on juicy cacti, taking slow bites, but also eat the leaves of low trees and bushes. Many iguanas do not drink at all, obtaining all the liquid they require from cacti, but those living near springs always drink regularly. Iguanas browse from 10 a.m., spend the midday heat in the shade, and forage again in the afternoon. After 4 p.m. they shelter in their burrows. The female lays several oblong, leathery eggs in holes dug in the ground. The young hatch after 50 days.

Galápagos Land Iguana

CUBAN GROUND IGUANA
Cyclura macleayi

The Cuban Ground Iguana is a massive iguana. It is 1.5 m long, but reaches a weight of 15 kg. It has a tall, sharp crest on its back, and long, pointed scales on its tail, which it uses to defend itself. It also has powerful claws. The home of this iguana is in Cuba, where it is a protected species. It is predominantly terrestrial, but young iguanas sometimes climb up low trees to rest and hunt for nestlings. Young iguanas feed mainly on insects. The adults are omnivorous, while older specimens prefer sweet fruits and shoots. Iguanas forage after sunrise, but seek cool shady places around midday. They sometimes bathe in shallow water.
The female lays 2 — 6 leathery eggs in a depression dug in the ground, and carefully buries the clutch, covering it with leaves. The young hatch after 50 days. They have alternate crosswise pale and dark stripes, and this protective coloration helps them to escape the attention of predatory birds seeking them among the tangle of vegetation in which they live.

CRESTED KEELED LIZARD
Leiocephalus carinatus

The crested keeled lizard *Leiocephalus carinatus* is common in Cuba. Its sturdy body reaches a length of 30 cm. It moves quickly and nimbly both on the ground and on walls, and it climbs tall cacti and bushes to catch insects. Large specimens occasionally catch small anoline lizards. When surprised, the lizard stops abruptly, raises the front of its body, and curls its tail into a spiral above its back to display the bright coloration at the base of the tail. At the same time it changes from a pale to a dark colour. In this way it often manages to deter its attacker, or use the moment of surprise to dart into its shelter. When the danger is over, it carefully pokes its head out of the hole, looks around warily, and then crawls out. It

Cuban Ground Iguana

Crested keeled lizard Leiocephalus carinatus

likes to bask in the sun in front of its shelter, but spends the midday heat in the shade.
The female lays about 12 eggs with leathery shells in a depression dug under a stone. The young hatch after 5 — 6 weeks. These lizards fall victim to many predators in the wild, especially birds and snakes.

TREE TRUNK ANOLE
Anolis porcatus

Anolis porcatus is the anole most commonly found in Cuba. It measures up to 25 cm. The male is larger, and more strongly coloured than the female. It is green above, but turns brown when irritated. It also changes colour when in hiding, completely blending with its background. The female is brownish, with a green or yellow stripe on her back.
This anole is found in parks, gardens, woods and bush-covered localities. It climbs trees and cacti, choosing a permanent site in a cleft between leaves or in a cavity, and crawling out to bask in the sun, often on a tree trunk. It hunts insects and their larvae on leaves, trunks and cacti, and sometimes itself falls prey to raptors.
The female lays her eggs into a hole which she digs out. The young hatch after 55 days and immediately look for their own territories and shelters.

Tree trunk anole Anolis porcatus

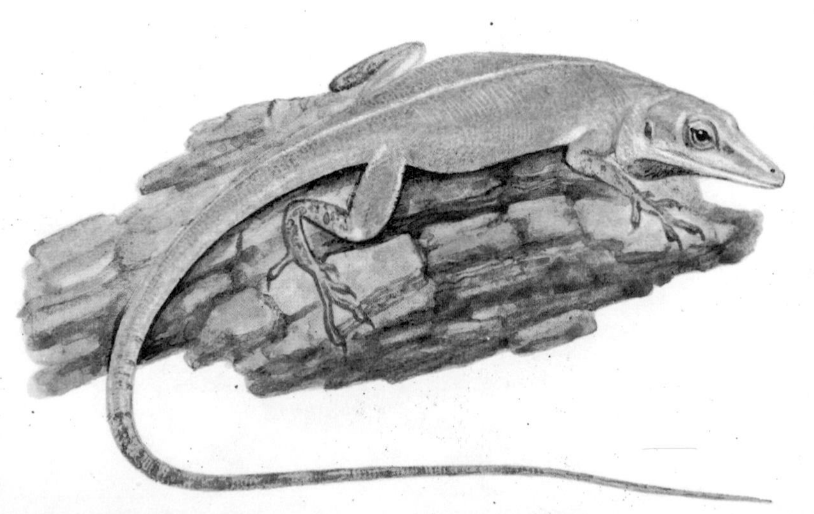

CUBAN TEIID
Ameiva auberi

The teiid *Ameiva auberi* is a slender lizard found in Cuba, where it inhabits localities with occasional bushes or thin woodland. It measures about 30 cm including the very long tail. This lizard is predominantly terrestrial, seeking suitable holes or digging its own burrows in sandy soil, though it may sometimes climb trees or bushes. On sunny days, teiids hunt insects and spiders. They are extremely agile, catching their prey with lightning speed. They swiftly hide in crevices or among stones when threatened by a bird of prey. During the midday heat, teiids rest in the shade. The female lays about 15 eggs in a depression which she digs for the purpose. The young hatch after about 50 days.
Fifteen other similar species are found in Central America, Mexico and the West Indies.

CUBAN BOA
Epicrates angulifer

The Cuban Boa was once hunted for its high-quality skin. It can reach a length of 4.5 m, but such gigantic specimens are now rare. This slender snake lives in localities where there are caves, spending much time in them, and even hibernating in them, although winter in Cuba is very mild. It also hunts bats in the caves, living almost exclusively on these creatures although it may catch small birds among the trees.

The female sometimes bears live young, but the tiny snakes can be born inside transparent coats which are easily broken out by the hatched young. Immediately after birth the young snakes open their mouths and make wild, jerking head movements to deter potential enemies. After 2 hours, they glide away to hide in crevices and cavities. Each litter contains up to 10 young. This snake is a protected species in Cuba.

Cuban teiid Ameiva auberi

GIANT TOAD
Bufo marinus

The Giant Toad is common in Central and South America, being particularly abundant in plantations, especially fields of sugar-cane. It is always a welcome visitor, because it catches insects, larvae and molluscs which damage the crops. It is strictly protected and has been introduced to new localities. Efforts have been made to acclimatize this toad in warmer regions of North America, for it has been calculated that a single Giant Toad can save up to 40 dollars worth of crops every year.
With a length of up to 25 cm, this is one of the largest toads in the world. In the breeding season, the males produce hoarse croaking sounds which can be heard several hundred metres away at night. The toads mate in water, where the female lays her eggs in jelly-like strings, up to 3 m long. The tadpoles hatch after 10 days and develop in the water for 80 days, undergoing metamorphosis and crawling out as tiny frogs which mature after 5 years. The Giant Toad is noted for its longevity — it lives to be almost 40 years old!
During the day the toads shelter in a damp spot under leaves, in holes, or under stones. At dusk, they begin to forage for insects, worms and molluscs, and the largest toads even catch small vertebrates. Like other related species, this toad has a warty skin containing many tiny venom glands. It also has special

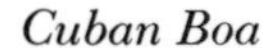

Cuban Boa

Giant Toad

in the rainy season. After about 6 weeks, tiny toads crawl out of the water in search of a moist place in which to live. At the beginning of the dry period, the toads dig out vertical burrows over 1 m deep in soft soil. They shelter in their underground chambers during the day, leaving them only after sunset to hunt small spiders, insects and larvae, molluscs and worms. However, it is possible to see this toad during the day without digging it out. All one has to do is to sing or produce a loud sound, and the toad will appear to find out what is going on. It is not known whether the toad is curious, or is just disturbed by the noise.

glands behind its ears which exude an irritant which gives rise to a burning sensation in the mouths of carnivores. This helps to protect it from predators.

CUBAN TOAD
Bufo empusus

The Cuban Toad is another amphibian found in Cuba. It measures about 10 cm and inhabits open localities covered with short grass and thin bushes. It occurs in large parks in the outskirts of Havana, near ditches full of water where the female lays her eggs

BROWN RECLUSE SPIDER
Loxosceles laeta

The brown recluse spider *Loxosceles laeta* lives in South America, mainly in Chile, Peru and Argentina. Despite its small size — the female measures 25 mm and the male is about a third smaller — its bite gives rise to acute poisoning. The venom breaks down the tissue around the bite, sometimes causing the area to swell up and turn black. In human victims sensitive to such poisons, treatment may take several months. As well as living in rocky sites this spider frequents human settlements, so people often fall victim to it. It spins large irregularly-shaped webs in the corners of windows, and feeds mainly on insects caught in the webs. The female lays 50 or more eggs in a casually woven sac, which hangs from the web.

SILVER ARGIOPE
Argiope argentata

The Silver Argiope inhabits tropical regions in South and Central America. This spider is found on the fringes of forests, in gardens and thickets, and among tall plants. The female is about 25 mm long, and the male only 10 mm. This spider is plentiful in some localities. It spins huge webs between branches or stems, and sits in them as it waits for prey, such as flies, butterflies and wasps to become enmeshed. The female lays 100 — 300 eggs enveloped in silken webs. The cocoon is hung between two grass stalks or on a twig. The hatched spiders stay together for a few days and then disperse.

Cuban Toad

Brown recluse spider Loxosceles laeta

Many other species of the genus *Argiope* are found in America.

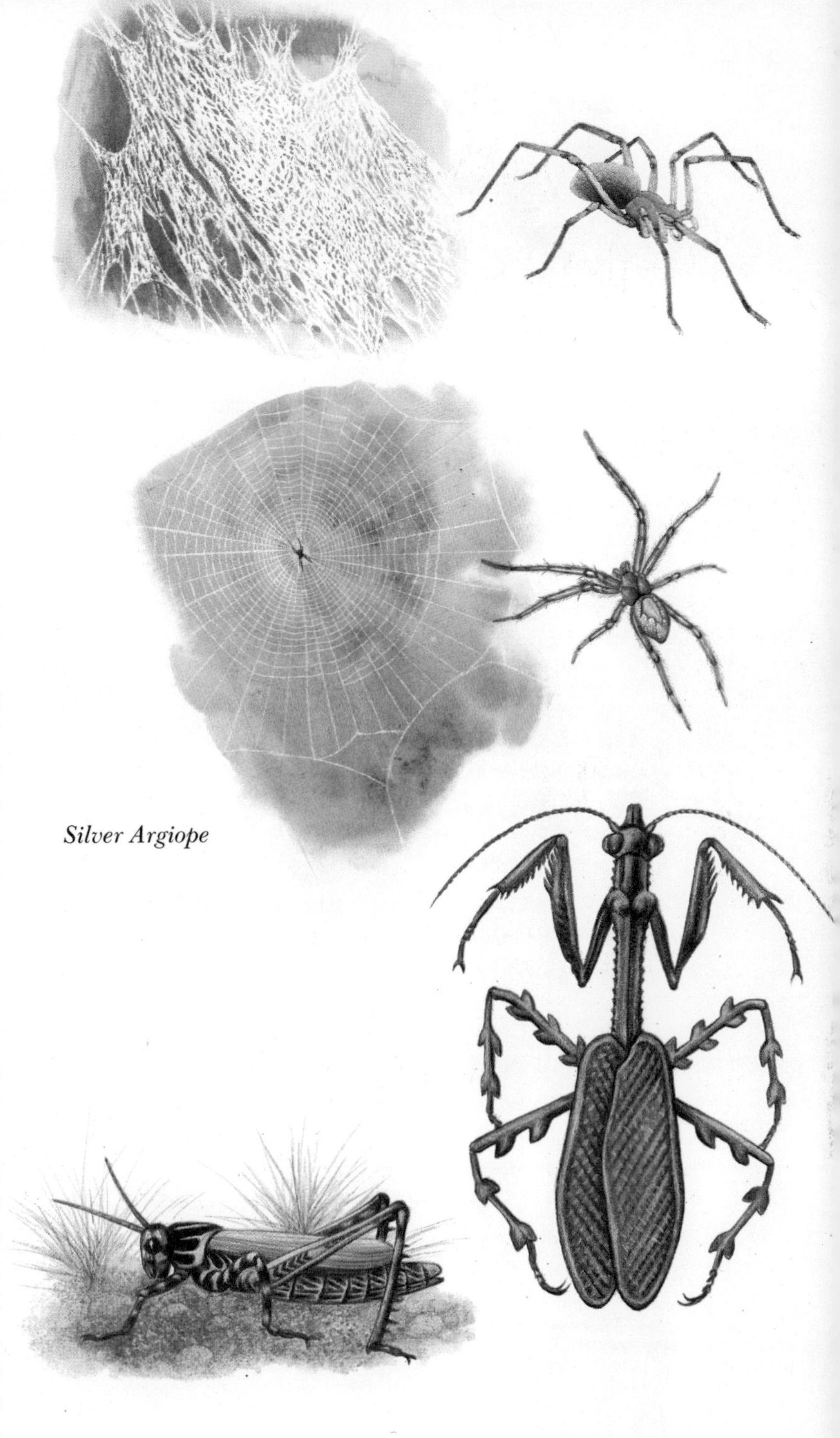

Silver Argiope

GRASSHOPPER
Zoniopoda omnicolor

The varicoloured grasshopper *Zoniopoda omnicolor* reaches a length of 35 mm, and is widespread in grasslands and bush-covered localities in South America. In the breeding season, the fertilized female's abdomen becomes much longer, and she exudes a mucous fluid. This secretion stiffens the walls of a depression which she digs out with her abdomen. She lays 30 eggs in the prepared hole.
She lays several batches in the course of a year, and when they have hatched, the nymphs crawl out of the hole and disperse. They moult for the first time at the age of one week, and shed their skins five more times before reaching the adult stage. Both nymphs and adults feed on grass and delicate leaves.

PRAYING MANTIS
Vates multilobata

The praying mantis *Vates multilobata* lives in tropical regions of South America. It has a markedly elongated thorax and conspicuous outgrowths on its legs. The female measures about 10 cm, but the male is only about half this size. This mantis frequents bushes and tall plants, its coloration and body shape enabling it to blend perfectly with its background. This mimicry makes it invisible to both prey and predators. When a butterfly, wasp or other insect alights nearby, the locust immediately extends its first pair of legs and grasps the prey between its femurs and tibias, which are covered with sharp spines. It devours the whole prey except for the tough indigestible parts.
In the mating season, the male seeks a female, and although they are poor fliers, they do fly at this time. The fertilized female lays severeal hundred eggs in a case 8 cm long, which she attaches to a twig or stem. The case hardens and protects the eggs from predators and from drying up. The nymphs hatch after a few weeks and are highly rapacious. They mature after a number of moults.

Praying mantis
Vates multilobata

Grasshopper
Zoniopoda omnicolor

Tarantula-hawk wasp Pepsis cinnabarina

TARANTULA-HAWK WASP
Pepsis cinnabarina

The tarantula-hawk wasp *Pepsis cinnabarina* is a robust insect which inhabits open, sandy country in Central America, Mexico, and the south-western part of the United States. It measures about 7 cm, and has long legs covered with tough bristles. It flies just above the ground among bushes in search of spiders, as large or larger than itself. Interestingly, the threatened spider usually offers no resistance. The wasp stings its prey in a nerve centre on the underside of the body, and the strong venom almost immediately paralyzes the spider without killing it. The wasp leaves the spider where it attacked it, which is often the spider's own nest, and lays an egg on the spider's body. The larva hatches out on the following day, and feeds on the paralyzed prey, at first avoiding its nerve centres. This ensures that the spider remains alive to provide a continuous store of fresh food. When the larva pupates, the involuntary host dies, and in due course a wasp emerges from the pupa.

BUTTERFLY
Dryas julia

Dryas julia is a butterfly of the family Heliconiidae. It has a wingspan of up to 10 cm. This butterfly occurs in bush-covered situations in southern Florida, Texas and Central America, ranging down to the north of South America. The female is paler than the male and clumsier in movement. The male is an excellent flier, often covering great distances. This butterfly repels many predators by means of a strong, foul smell. The female lays her eggs on passion-flowers, the leaves providing food for the caterpillars. These are covered with sparse, long, thorny hairs. They pupate after several weeks, usually hanging upside down.

DIADEM BUTTERFLY
Hypolimnas misippus

The Diadem Butterfly is common in the West Indies, and is found in some localities in Florida. It was probably brought to the American tropics on slave ships sailing from their African homeland. It also occurs in India, Pakistan and Australia. The male and female differ both in size and coloration, the two sexes being at first described as two species. The fe-

Butterfly Dryas julia

Diadem Butterfly

♂

♀

male has a wingspan of about 8 cm and in her appearance imitates other butterflies, which repel insectivores with their fetid smell. The male has a wingspan of about 6 cm. Logically, the female should be less susceptible to predation, but in fact both partners have equal chances of survival. This is because the female has poor powers of flight, and easily falls prey to birds. The male, on the other hand, is capable of sustained and fast flight, and so escapes his natural enemies. The caterpillars live predominantly on plants of the nettle family which grow in bush-covered localities.

Moth Urania leilus

MOTH
Urania leilus

The moth *Urania leilus* is one of the most beautiful of day-flying moths. It lives in bushy woodland in Trinidad, Guyana and Venezuela. It has a wingspan of about 8 cm. It flies above bushes, or among the trees, and in the evening, it settles on tree trunks to sleep. These moths leave their native land in large groups, to undertake a long journey south, as far as Argentina. They are excellent fliers, daily covering up to fifty kilometres. The trees on which they sleep swarm with these beautiful moths. It is not yet known why they set forth on these long trips, but overpopulation would seem to be a possible reason.

Moth Urania fulgens

Swallowtail butterfly Papilio pausanias

MOTH
Urania fulgens

The moth *Urania fulgens,* which has a wingspan of about 6 cm, is distributed from Mexico to Ecuador. It is also migratory, large numbers flying to the south or south-east in June and July. At this time they are seen in abundance in the vicinity of the Panama Canal. For several days, thousands of these beautiful moths fly above it, often alighting on ships sailing through. Some moths fly eastwards, to the sea, and these perish. This has the effect of reducing overpopulation. In March and April, the moths return to their home in the north.

Butterfly Phoebis avellaneda

SWALLOWTAIL BUTTERFLY
Papilio pausanias

The swallowtail butterfly *Papilio pausanias* is native to the tropical regions of South America. It has a wingspan of about 10 cm, and the hind wings have sword-like prominences. On sunny days the swallowtail flies among the bushes and over the grass at the edges of woods, and in parks and gardens. It glides rapidly, usually low above the ground. It seeks flowering plants, settling on the flowers and sucking sweet nectar with its long proboscis. Early in the morning, it drinks water from humid places on the ground. The caterpillar, like those of most swallowtails, is able to project a red forked tubercle from the top of the thorax, to deter approaching predators. It also exudes a fetid fluid which repels insectivorous predators.

BUTTERFLY
Phoebis avellaneda

The butterfly *Phoebis avellaneda* has striking yellow and carmine coloration and a wingspan of about 7 cm. It frequents grassy or bush-covered localities in Cuba and is abundant in some localities. It is a good flier and often visits gardens where it settles on flowering plants. Early in the morning, scores of these butterflies can be seen as they drink from puddles.

AQUATIC HABITATS OF SOUTH AND CENTRAL AMERICA

Let us now visit the lakes, marshes and rivers of a part of the world which has not yet been fully explored. It is a world rich in noise, colour and peril. It is the home of the largest crocodile, and of thousands of tetras which find a second home in our aquaria. It is also the habitat of their dangerous relative, the piranha. But no one can estimate how many more species of fishes are still waiting to be discovered and find their place in books such as this, or how many beautiful birds and other remarkable animals live there.

The world of water is as different from that of deserts and mountains as another continent. Vast river basins are situated in luxuriant rainforests dripping with water. Of the aquatic fauna only fishes and some insects live beneath the water. The other animals, especially water and wading birds, live in association with terrestrial creatures, and the borders of their two worlds overlap and blend. There are moments, when the line between water and dry land becomes a matter of life and death. Such a time is in the rainy season, when the rivers overflow their banks and the life-giving water turns into a force of destruction. It washes away banks, uproots trees, and floods vast savannahs. Nowhere in the world can one see the devastating effect of water on such immense scale as in the vicinity of the South American rivers.

Who has not heard of the Amazon, the largest river on our planet. The Nile may be 191 km longer, but the Amazon with its tributary the Ucayali measures 7 025 km, surpassing the Nile and the Kagera put together. No river can match the Amazon in the quantity of water it carries.

It flows at a rate of 120 000 cubic metres per second, which is the equivalent of fifty European rivers put together. Its basin covers about a third of the continent of South America, an area of some 7 million square kilometres. The Amazon rises 150 km from the Pacific coast, in the mountains of Peru and zigzags eastwards across the continent. As it flows it is joined by hundreds of tributaries and five large rivers before it finally hurls its muddy yellowish water into the blue Atlantic. Three hundred kilometres beyond the delta, the ocean still runs yellow under the massive impact of the Amazon. The enormous river is used by ocean liners travelling thousands of kilometres upstream, into the heart of the continent, and large vessels can reach the Peruvian town of Iquitos beneath the foothills of the Andes.

The Rio Negro is the largest tributary of the Amazon. As its name — Black River — suggests, its water is dark coloured by peat from the rainforests. The two rivers meet at Manaus, but for many kilometres below the town each keeps its own colour.

The Orinoco, called 'Father' by the native population, is another great river in the north of the South American continent. It is a source of food for the Indians who fish in it, catch freshwater turtles, and collect millions of eggs laid by tortoises in the sand of islands in the river. It is also the home of the American Crocodile, which can reach a length of 7.5 m.

The banks of the Orinoco, especially the mangrove-covered islands of the muddy delta where it approaches the sea near Trinidad, are the home of aquatic, wading and predatory birds. The Orinoco system is connected to the Rio Negro by the Casiquiare, a river 220 km long in the southern tip of Venezuela. As the Orinoco flows to the west, some of its water runs into the Casiquiare, which turns south along the border between Colombia and Brazil before joining the Rio Negro, which in turn flows into the Amazon. A channel of this kind linking European rivers is the dream of navigators, merchants and engineers, but South America obtains no economic benefit from the Casiquiare. It is virtually inaccessible, is overgrown with vegetation, and is scarcely navigable. Worse, its slow-moving warm water is the breeding place of millions of mosquitoes, which plague the region.

The South American rivers are also outstanding for their spectacular waterfalls, Iguaçú Falls, on the border between Brazil and Argentina being the most famous. These falls are 4 km wide and 85 m high. Angel Falls in Venezuela, with a drop of 1054 m, is the highest waterfall in the world, and the Potaro River in Guyana features famous falls, 222 m high.

Lakes, vast stretches of marshland and lagoons occupy a special place in the world of water in South and Central America. The lagoons are large shallow lakes separated from the sea by sand bars, and containing very warm brackish water — a paradise for both animals and naturalists.

Mangroves typify the lagoon vegetation, the tide regularly revealing the stiltlike roots on which the trees stand as if on a pedestal. The characteristic picture of the tropics which they create is further enhanced, especially in Cuba and Yucatan, by flocks of beautiful flamingoes, herons, ibises, and many other birds. Some lagoons which are connected to the sea are visited by small fishes, especially rays, which bask in the warm water on the sandy bottom. The lagoons are also inhabited by crocodiles, frogs and crabs of all colours.

The lakes of the high mountains provide a peaceful and majestic spectacle, although they teem with life. With an area of 14 350 square kilometres Lake Maracaibo in Venezuela is the largest lake in the continent, but Lake Titicaca, situated at a height of 3 812 m on the border between Bolivia and Peru, is more famous. Its shores feature both the ruins of the most ancient city on the American continent, and the modern structures of La Paz in Bolivia. The lake is the home of 20 species of fish of the subfamily Orestiidae which exist nowhere else.

Shallow, glacier-fed salt water lakes are situated in the mountains of Peru, Bolivia, Chile and Argentina, at heights above 4 000 m. Here three species of flamingoes, feeding on red algae are to be found. They are the Chilean Flamingo, the Andean Flamingo, and James's Flamingo. All three species are threatened by extinction, for the local Indians collect their eggs and catch young birds.

WATER OPOSSUM or YAPOK
Chironectes minimus

The Water Opossum or Yapok is widespread beside freshwater lakes and streams of Central and South America. It is about 40 cm long, and is the only marsupial adapted to life in the water. It has thick fur, and webbed hind feet which enable it to move quickly in the water, where it catches fishes, amphibians and crabs. It carries small prey in its facial pouches and holds large prey in its teeth, eating the food on land. When eating a crab it first crushes the tough carapace with its strong teeth, and then picks out the soft body. Heaps containing the remains of fishes and crustaceans mark the Yapok's habitat.
The Yapok shelters in tree cavities or holes in the banks of the streams. It sleeps during the day, sometimes basking in the morning sun. It is a fast runner, and when disturbed, it quickly disappears into its hole or escapes under water. After a gestation period of 12 — 13 days, the female gives birth to about 6 young, born naked and blind. They continue to grow in their mother's pouch. She is able to dive even with her offspring inside, for she can close the pouch by means of special muscles, making it watertight.

THICK-TAILED OPOSSUM
Lutreolina crassicaudata

The Thick-tailed Opossum is common in marshland and on river banks of the eastern part of tropical South America. Its total length is about 65 cm. During the day, it shelters in holes in trees near to the ground, or among the roots. It forages after sunset, hunting fishes in shallow water, frogs and crabs on the shore, and sometimes taking the eggs of water birds. Young opossums feed on insects. In the mating season the male and female stay together for a few days, but for most of the year opossums are solitary. The female gives birth to 6 — 10 naked and blind young after a 13 day gestation period. For 8 weeks the young grow in their mother's pouch, attached to her nipples and being nourished by her milk. When they come out they are carried on their mother's back. At the age of 10 weeks young opossums begin to crawl, and they become independent after 3 — 4 months. They are sexually mature after 6 months.

MEXICAN BULLDOG BAT
Noctilio leporinus

The Mexican Bulldog Bat is found throughout the tropical regions of South America, and is quite abundant in some localities. It has a wingspan of about 45 cm, and is one of the smaller species. It has long legs with elongated toes and strong hooked claws. In some localities this bat feeds largely on fish. It flies out of its shelter in a tree hollow after sunset, and skims over the surface of the river or lake, dipping its feet into the water, and catching tiny fishes swimming near the surface. The bat then flies up and eats its prey on the wing. These bats also hunt flying

Water Opossum or Yapok

Thick-tailed Opossum

insects, especially on the edges of forests and along rivers. After a gestation period lasting 70 days, the female bears a single young which she carries clinging to the hair on her chest as she flies.

Mexican Bulldog Bat

GIANT OTTER
Pteronura brasiliensis

The Giant Otter is indeed a giant among its relatives. It is up to 2.2 m long including a 70 cm long tail, and it weighs up to 25 kg. This species occurs from Venezuela and Guyana south to Uruguay and northern Argentina, but it no longer exists in many places where it was once found. It builds spacious burrows in the banks of rivers or lakes. The otter is chiefly diurnal, but often hunts at dusk. In localities where it is hunted, it leaves its shelter only at night. It catches its prey in water, where it swims fast and skilfully, and takes its catch to the shore to eat it. Frogs and aquatic birds are sometimes caught as well.
The Giant Otter mates in the water, and after a gestation period of 60 — 65 days, the female bears 1 — 2 young. These are born naked and blind, gaining their sight after 28 — 35 days. Shortly afterwards, they begin to play in front of the burrow, but they do not learn to swim until they are 3 months old, by which time they have been weaned. By the time they are 4 months old they can swim and dive skilfully and are beginning to hunt. They mature at 2 years and live to be 10 years old. Some females have two litters in a year.

CAPYBARA
Hydrochoerus hydrochaeris

The Capybara is the largest of all the rodents. It is over 50 cm tall, 1.3 m long, and weighs over 50 kg. It inhabits the tropical regions of South America from Panama to north-western Argentina, where it roams in bands in dense vegetation around rivers and lakes. It has massive, broad incisors, webbed feet and hooflike claws. It has three toes on its hind limbs, and four on its forelimbs. The sexes are difficult to distinguish, because the sexual organs are hidden in folds of skin. In the mating season capybaras exude a liquid from a scent gland situated on the upper part of the mouth. This attracts females over a large distance so a male can gather a small herd. During the day, capybaras shelter in the vegetation seeking refuge in the river when in danger. They forage at dusk, feeding predominantly on green plants, bark and fallen fruit. They often travel a kilometre or more from the river in their search for food. They move clumsily on land but are in their element in water, being excellent swimmers.
Once a year, after a gestation period of 119 — 126 days, the female gives birth to 2 — 8 young amid thick vegetation. Capybaras are born well-developed,

Giant Otter

Capybara

covered in fur, and with open eyes. They begin to run around on the following day and nibble grass. The Capybara's average lifespan is 12 years. Its chief enemies are jaguars, and men who hunt it for its palatable meat.

NUTRIA
Myocastor coypus

The Nutria is a well-built South American rodent, weighing up to 9 kg. The body measures 40 — 80 cm and the tail 40 — 50 cm. It frequents river banks in which it digs burrows 1 — 6 m long. Where the banks are stony, nutrias build nests among reed beds. They bend the reed stalks, to form 'baskets' in which to shelter. The Nutria seeks its food mainly in the water. It is a good swimmer and dives to the bottom to dig out stems and roots of aquatic plants. It also feeds on grass and leaves, gnaws the bark on young branches, and eats fallen fruit. It can remain submerged for up to 5 minutes so it easily escapes its enemies.

Nutria

After a gestation period of 100 — 150 days (130 on average), the female gives birth to 1 — 10 young in the burrow. These are born with open eyes and are covered with thick fur. They are suckled for 6 — 8 weeks, but after a few days they begin to eat green food. They mature after 5 — 6 months, and a female may have 2 — 3 litters a year. Despite its fecundity, the Nutria has been exterminated in many places, having been hunted for its valuable fur. It has, however, been introduced in other areas. At the turn of the century, nutrias were introduced in the United States and in Europe, and they are now reared on a large scale on farms.

BRAZILIAN TAPIR
Tapirus terrestris

The Brazilian Tapir is found in four subspecies from Mexico south to northern Argentina. It is up to 1 m tall at the shoulder, and about 2 m long, females being slightly larger. The tapir is characterized by its elongated upper lip which resembles a short trunk. It dwells in marshland and swamps near rivers, where it likes to splash and wallow. It is an excellent swimmer and diver. During the day it rests, hidden in thickets or reed beds. It leaves its shelter at dusk and always uses the same well-trodden paths. First it enters the water where it often remains for hours, with only its head showing. Later it browses on the leaves of bushes, grazes or peels the bark from trees. It also feeds in fields of maize.

Brazilian Tapir

Tapirs are solitary except in the mating season, when the male seeks a female and stays with her for a short time. Immature animals roam together in small groups. After a gestation period of 390 — 405 days, the female gives birth to a single young which is dark, with yellow and white spots and lengthwise stripes. This seemingly over-colourful coat in fact allows the young tapir to blend with its background and so escape the attention of predators. A young tapir can see at birth and is able to walk within a few hours. For the first few days it spends much of its time asleep while its mother guards it, bravely chasing away even carnivores. Only a large and experienced jaguar dares to attack an adult tapir. The infant is very playful and runs around its mother. It loses the pattern on its coat when it is about 8 months old. Young females are mature at the age of 3 — 4 years, and males after 5 years. They live to be about 30 years old.

HUMBOLDT PENGUIN
Spheniscus humboldti

The Humboldt Penguin is a medium-sized species, about 70 cm tall. It is distributed along the western coast of South America, from Chile to Peru, and is very abundant in some places. It nests and rears its offspring almost throughout the year along the shores of rocky islands. Penguins are gregarious, living and breeding in large colonies of up to several

Humboldt Penguin

Red-billed Tropicbird

thousands. Pairs dig burrows in soft soil, or occupy suitable cavities or crevices. The nest is sparsely lined with pieces of wood, pebbles, fish bones and scales. The female usually lays 2 whitish eggs and incubates them alternately with the male. The young hatch after 38 — 43 days, and are covered with thick down. The hungry nestlings push their beaks into those of the adults, and feed on regurgitated food. Young penguins grow very quickly and leave the nest when they are 3 months old. They immediately begin to swim and soon are independent. Before entering the water, they swallow a few stones. This habit was once believed to facilitate diving, but the stones have a different purpose. They help to crush food in the stomach.

Young penguins roam in the neighbourhood of their home, and return to the nesting ground after 3 years, when they are adult and able to have their own families. The Humboldt Penguin swims under water, using its paddle-shaped wings as propellers, and its feet as the helm. It can reach a speed of 30 — 40 km per hour, easily catching fishes, which are its staple food. It also feeds on molluscs and crustaceans.

RED-BILLED TROPICBIRD
Phaëton aethereus

The Red-billed Tropicbird spends its life flying high above the sea, often hundreds of miles from the nearest land. Only in the nesting season is this heavily built bird found on land, on the shores of tropical South and Central America. It is up to 1 m long, the tail feathers being elongated, and the short feet having four webbed toes. It has a wingspan of about 110 cm. Tropicbirds have keen eyesight. Spying their prey of fishes or molluscs from a great height, they plummet into the water and seize it. They often accompany ships, and with shrieking cries snatch food thrown into the sea.

Nesting takes place in any season. Tropicbirds do not build an actual nest. The female lays a single egg in a sandy depression in the ground, or on a rocky ledge. The egg is reddish, with purple and brown spots. It is incubated for 28 days by both parents, who proceed to look after the young chick and feed it for 70 — 85 days, after which time the fledgling leaves the nest.

MAGNIFICENT FRIGATEBIRD
Fregata magnificens

The Magnificent Frigatebird is an outstanding flier, and can remain airborne for hours above the ocean. It inhabits the shores of tropical South and Central America and reaches a length of 94 — 114 cm. Its wingspan is 225 cm. Since its feet are extremely short, it never settles on the ground, and only excep-

Magnificent Frigatebird

tionally alights on water. It roosts always on tall trees or bushes from which it can easily take off. Frigatebirds feed only in flight, snatching their food from the surface of the sea. They often pursue other birds, such as terns, harassing them and forcing them to drop or disgorge their prey, which the frigatebirds catch in the air and swallow immediately. They also catch flying fishes as they skim above the water. In late afternoon, the birds begin to search for trees and bushes in which to spend the night.

In the courtship season, the male inflates a large, naked, red sac on his throat to attract a mate. The selected female of his choice perches on a branch, or on a nest of dry branches next to the male, and the birds face each other, rattle their bills and flap their wings. Frigatebirds nest in colonies, nesting time varying with the locality. The female lays a single white egg in a nest in a tall bush and both partners incubate it for 40 — 50 days. The young chick is naked at first, but soon grows thick down. It stays in the nest for 4 — 5 months, and even after leaving the nest, it is fed for a further 2 — 6 months.

Guanay Cormorant

GUANAY CORMORANT
Phalacrocorax bougainvillei

The Guanay Cormorant lives on rocky islands and cliffs along the Peruvian coast. The islands, 30 — 50 km from the mainland, suffer rainfall for only two days a year, so they provide ideal nesting conditions. This cormorant reaches a length of 65 cm, has a wingspan of 100 cm, and lives in colonies. The islands are small, so the birds are crowded together, as densely as three pairs to one square metre. On one island alone, numbers in excess of one million have been counted, and the total population is estimated at 20 millions, the largest existing bird population. Other species of birds nest on the islands, but cormorants account for 85 per cent of the total. Why have so many birds, all feeding on fish, settled in this one place? The reason is simple. The shores of these islands are washed by the cold Humboldt Current, which provides an excellent environment for enormous quantities of anchovies, on which cormorants feed. The birds swim on the surface and dive to catch their prey.

The presence of this colony along the Peruvian coast was known to the Incas. They declared it a protected area, and ensured that the nesting grounds of these birds were inaccessible. Today these birds are still protected, regardless of the fact that they consume huge quantities of fish. This is because their excrement, rich in phosphorus and other minerals, is the source of guano, one of the best natural fertilizers. Over the centuries, deposits of guano several metres high formed on the islands. Nevertheless, these supplies were soon exhausted and nowadays the new layer is scraped off annually after the nesting season. Since just one colony daily produces up to 1 000 tonnes of guano, the amount which is collected and shipped away in bags, makes this a very profitable enterprise. Nesting can take place in any season, but most birds nest in December and January. The females lay their eggs on the ground, in a nest lined with a few feathers. The clutch of 3 — 5 bluish-green eggs is incubated alternately by both sexes for 29 — 30 days. The young hatch one at a time over a period of several days because the adults start incubating as soon as the first egg is laid. The parents feed their offspring on regurgitated fish, later bringing fresh fish in their throat sacs. After 8 weeks the young birds

Anhinga, Snake-bird or Darter

Peruvian Booby

leave the nest. The flocks then travel along the Peruvian and Chilean coast, flying as far as Colombia.

ANHINGA, SNAKE-BIRD or DARTER
Anhinga anhinga

The Anhinga, Snake-bird or Darter, is related to the cormorants. It has an incredibly long neck and a long, thin, pointed bill. It is about 60 cm long and has a wingspan of 120 cm. This bird frequents inland aquatic habitats ranging from the southern United States to Argentina, especially localities overgrown with tall bushes and sparse woodland. It lives on fish, speared under water with its pointed bill. The nest is built of sticks, in branches of trees or bushes, preferably over the water. The female lays 3 — 6 blue-green eggs, which are incubated by both partners for 25 — 28 days.

PERUVIAN BOOBY
Sula variegata

The Peruvian Booby is another sea bird which produces the valuable fertilizer guano. It nests on islands along the Peruvian coast, where its colonies form about 10 per cent of the total population of local birds. This represents some 1.5 million boobies. This bird is 65 cm long and has a wingspan of 150 cm. Pairs use seaweed to make their nests on rocky cliffs and platforms. The female lays 1 — 3 blue-green eggs and both partners share in incubation for 40 — 44 days.

BROWN PELICAN
Pelecanus occidentalis

The Brown Pelican is a common inhabitant of coastal regions from the southern United States to Peru, and also occurs in the West Indies. It reaches a length of 1 m and has a wingspan of 225 cm. It frequents lagoons or quiet bays along the coast where it searches for fish. It does not swim on the surface like other pelicans, but flies as high as 20 m above the water, and can hover in one spot by fluttering its wings. When it sees a fish swimming near the surface, it dives headlong down with half-closed wings, disappears under the water, often for several seconds, and then swims up with its prey. Then it

takes to the air again and slowly glides in search of another fish.
Like other species of pelicans, brown pelicans nest in colonies of several hundreds. The nest is built on rocky cliffs or among the branches of tall coastal bushes, especially mangroves, and is made of sticks, reeds and branches. The female lays 2 — 4 bluish-white eggs and takes turns with the male to sit on them for one month. The offspring are taken care of by both parents, who feed them on fishes which the young pick out of the adults' throat sacs. After 12 weeks, young pelicans are capable of flight, and can fend for themselves after 15 weeks. They mature after 3 — 4 years. The Brown Pelican is mute, producing only clacking sounds with its bill, or hissing.

SNOWY EGRET
Egretta thula

The Snowy Egret is distributed from the southern United States to Chile and Argentina. It is 55 — 65 cm long. It dwells in lagoons, river deltas and coastal swamps, and frequents rice paddies. In autumn, flocks of egrets migrate from the north southwards, often in the company of other wading birds. They nest in colonies, at times which vary with the locality. The pairs build their nests in mangroves or other trees growing in swamps. Over 20 nests of twigs, reed stems and other plant material, may be built in one tree. Both partners incubate the clutch of 2 — 5 greenish-blue eggs for 24 days. They start incubating from the time the first egg is laid so the young hatch gradually. The parents bring food in their throat sacs, and the nestlings pick it out. At the age of 6 weeks, young egrets are capable of flight and begin to hunt, feeding on tiny fishes, molluscs, crustaceans and insects.

YELLOW-CROWNED NIGHT HERON
Nyctanassa violacea

The Yellow-crowned Night Heron is distributed from the central part of the United States to Brazil and Peru. In autumn it migrates from the north southwards, and returns in spring. It is 56 — 71 cm long, has a wingspan of 110 cm, and produces

Brown Pelican

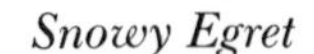

Snowy Egret

a croaking call. It frequents coastal areas, rocky islands and inland marshes, and except in the nesting season, it travels in flocks or singly. The birds often stand in shallow water as the tide ebbs to catch crustaceans, particularly crabs. In the nesting season, the pairs gather in colonies and build nests of sticks in mangroves. The female lays 2 — 6 bluish-green eggs and sits on them alternately with the male for 22 days. The young, who are cared for by both parents, leave the nest after 5 weeks, but perch at first on nearby branches. They are capable of flight after 8 weeks, at which time they begin to fend for themselves.

Yellow-crowned Night Heron

BOATBILL HERON
Cochlearius cochlearius

The Boatbill Heron is 60 cm long, and has a broad, scoop-like bill. It lives in freshwater mangrove swamps, ranging from central Mexico to Brazil and northern Argentina. During the day, it sits motionless among the branches, but it becomes active after sunset and hunts fishes, frogs and crustaceans in shallow water, and small mammals on land. In the nesting season, it lives in pairs or in small colonies. The nest is made of twigs and is situated in mangrove trees, usually directly above the water. The female lays 2 — 4 pale blue eggs and shares incubation with the male for 25 days. The young are looked after for 7 weeks, until they can fly. The Boatbill Heron produces harsh croaking sounds similar to those of frogs, and it often clacks or rattles its bill.

Boatbill Heron

WOOD STORK
Mycteria americana

The Wood Stork is a well-built wading bird about 1 m tall, and with a wingspan of 165 cm. It inhabits a wide zone from Florida to Argentina and eastern Peru, where it frequents swamps, lakes, lagoons, and coastal sites. It lives in flocks and forms colonies in the nesting season. The pairs build nests on mangrove branches where the female lays 3 — 5 white eggs. Both partners share in incubation, which lasts 28 — 32 days, and in parental duties. Young storks leave the nest at the age of 8 weeks and become independent after another 3 weeks. The birds prey on fishes, amphibians, reptiles and insects, caught in shallow water or on land. This stork is voiceless, but in the nesting season it produces hissing sounds and clacks its bill.

JABIRU
Jabiru mycteria

The Jabiru covers the tropical region from Central America to northern Argentina and eastern Peru. It is a sturdily built wading bird about 140 cm tall, and it has a stout bill. It spends the entire year near rivers, lakes or swamps, and in Brazil it is also resident in humid grassland. The birds wade singly or in pairs through shallow water or grass, searching for prey. They hunt mainly for fishes, molluscs, crustaceans, lizards and snakes, and occasionally catch small mammals. The nest is built in a tall tree, or on a rocky platform overlooking a river. The large construction is woven from branches and is thickly lined with grass, pieces of hide and mammal hair. The female lays 2 — 4 greyish or yellowish-white eggs. Both partners share in incubation for 35 days, and then both feed the young, which leave the nest when they are 11 — 12 weeks old.

MAGUARI STORK
Euxenura maguari

The Maguari Stork is found in the vast marshlands and along the rivers of tropical South America between Guyana and north-eastern Argentina. It is a wading bird, about 130 cm tall. It is resident in most territories, only the Argentinian population migrating north after the nesting season. Storks live in pairs not only during the year, but over a period of many years. They forage for food in shallow water, on banks, or in damp grassy localities, hunting for small vertebrates, and less often for insects. The nest is made from sticks, stems and leaves. It is a large construction, situated on a branch near to the tree trunk, and the same nest is used for many years. The female lays 3 — 5 whitish eggs, which are incubated by both partners for 32 days, though only the female sits at night. The chicks are fed for two months on the nest, and then they learn to fly and hunt, becoming independent after another three weeks.

Wood Stork

Jabiru

Maguari Stork

White or American Ibis

a wingspan of 95 cm. It is distributed from the southernmost regions of the United States to the north-eastern part of South America, and is common in the West Indies. This ibis inhabits the banks of rivers, lakes, and brackish lagoons. It hunts crabs, fishes and amphibians in shallow water, and occasionally catches small snakes on land. Nesting colonies are made up of several hundred pairs. The nest is built of twigs, mostly on mangroves or other tall trees by the water. The nesting material is collected mainly by the male, while the female does the building. She usually lays 3 bluish or greenish eggs with brown spots, and the clutch is incubated by both parents for 22 days. Both adults also take care of their offspring. The young birds which are a greyish colour, stay in the nest for 5 — 6 weeks, until they are capable of flight.

WHITE or AMERICAN IBIS
Eudocimus albus

The White or American Ibis often lives in the company of herons and scarlet ibises, and it sometimes interbreeds with them. It is about 60 cm tall, and has

SCARLET IBIS
Eudocimus ruber

The Scarlet Ibis has a discontinuous range of distribution, being found along the north-eastern coast of South America, and occasionally occurring in south-eastern Brazil. It is 52 — 68 cm long and has a wingspan of 95 cm. It forms large colonies, and flocks of scarlet ibises can be seen feeding on small fishes, amphibians, molluscs, crustaceans and insects in areas of shallow water. Nesting takes place in the season of tropical rains. Each colony numbers several hundred birds, and the nests are so close together, they sometimes touch. The female lays 2 bluish or greenish eggs with brown spots. Both partners incubate the clutch for 21 days and then feed the chicks. The young birds leave the nest after 5 weeks, at which time they are brown in colour. The Scarlet Ibis produces grating, three-syllabled sounds.

Scarlet Ibis

ROSEATE SPOONBILL
Ajaja ajaja

The Roseate Spoonbill is widespread in extensive lagoons near the coast, in marshes and in river deltas, throughout the whole tropical region of South America. It occasionally occurs in the southern United States. It reaches a length of 80 cm and has a wingspan of 130 cm. This bird usually lives in small groups, often in the company of other wading birds. It feeds in shallow water, walking with its half opened bill beneath the water. It moves its head from side to side as it walks, and snaps up any small fishes, crustaceans, worms or insects it encounters. Sometimes it eats the green leaves of water plants. This bird flies fast and straight, with its neck stretched forward, often covering many kilometres to reach its feeding grounds.

In the nesting season, spoonbills gather in small colonies on the marshes or on small islands, and build their nests in trees or bushes, or occasionally on the ground. The large nest is woven from branches, and is lined with fine leaves. The female lays 2 — 3 brown-spotted white eggs, and incubates them with her partner for 23 — 24 days. Young spoonbills leave the nest after 5 — 6 weeks, and become independent after another 2 weeks.

Roseate Spoonbill

GREATER FLAMINGO
Phoenicopterus ruber ruber

A century ago, over 50 huge colonies of the Greater Flamingo existed in America. There are only 10 nesting grounds in the world today, and the number of birds residing in them is steadily declining. Larger colonies have however been preserved in Cuba and in the state of Yucatan in Mexico. The total population of these beautifully coloured birds is currently estimated at 21 500, while only a few years ago there were 95 000 of them. The Greater Flamingo reaches a height of 150 cm and has a wingspan of 140 cm. It inhabits lagoons in Cuba, the Bahamas, Hispaniola, Yucatan, and in the north of South America.

It is a very shy and wary bird, and is especially sensitive to any kind of disturbance in the nesting season. A colony may move away from a place where the essential peace and quiet is lacking. The female builds a nest of mud, sand and decaying leaves on the shore or in shallow water. She piles up her building materials and then tramples them into shape, finally making a depression in the top of the 30 — 50 cm tall chimney-like construction. Here she lays a single egg, and incubates it for one month usually alone, though the male has been seen to take part. Flamingoes sit on their nests in the same way as other birds, with their long feet folded underneath them. The newly hatched chick has short feet and

Greater Flamingo

Chilean Flamingo

Andean Flamingo

a straight bill, and looks like a duckling, but after two weeks, its bill begins to curve and its feet grow bigger. The young chick stays on the nest for only 4 days, by which time it is strong enough to gather with other young birds to form flocks. They splash in shallow water, and the adults feed them with a special liquid mush. The parents always recognize their own young, and never feed other chicks. At the age of 3 weeks, the young are able to obtain their own food. The bill is held under water, bent at an angle, so that the upper part is beneath the lower part. The quickly moving thick tongue works like a pump, and the tip of the half-open bill sucks in water containing crustaceans, insects and plant debris. The water streams to the base of the bill and is filtered through a network of long, fine grooves along the sides of the bill. These seive out the tiny water animals and the flamingo swallows them.
Young birds begin to take to the wing after 11 weeks, and become excellent fliers within a very short time. Soon they are able to cover long distances, and seek their food far from the colonies.

CHILEAN FLAMINGO
Phoenicopterus ruber chilensis

The Chilean Flamingo is only 120 cm tall. It is easily recognized because it has grey-green feet with distinctly red joints. It is common in Peru, Bolivia and in the Chilean Andes, living in salt-water mountain lakes at a height of 4 000 m. It often frequents places virtually inaccessible to man. In winter, the huge flocks, often numbering several thousands, migrate from the alpine elevations to the lowlands, and especially to coastal areas. The Chilean Flamingo feeds on tiny crustaceans, insect larvae and algae. Its habits are similar to those of its relative, the Greater Flamingo.

ANDEAN FLAMINGO
Phoenicoparrus andinus

The Andean Flamingo is about 120 cm tall and inhabits south-western Peru, western Bolivia and northern parts of Chile and Argentina. It nests beside salt-water lakes at heights above 4 000 m, where the climate is severe. Even in summer, the nights are very cold, with the temperature falling to below freezing point, while at noon it climbs to 25° C. Both adults and young have to withstand these great variations, so why do these birds rest in such localities? Perhaps because in these saline mountain lakes, the birds are able to find sufficient suitable food. They feed on red algae, which are present here in such quantities that the water takes on the colour of blood. It also contains many small aquatic invertebrates.

Despite the remoteness of its nesting sites, the Andean Flamingo is not safe from man. The Indians collect both eggs and chicks and catch the adult birds during the moulting period, when for a few days they are incapable of flight. Very often whole colonies of these beautiful birds are quickly destroyed.

Little is known of the nesting habits of this species, and still less is known about its activities or whereabouts during the remainder of the year, for these birds have never been observed outside the nesting season.

James's Flamingo

JAMES'S FLAMINGO

Phoenicoparrus jamesi

James's Flamingo is one of the most beautifully coloured flamingoes. It is about 1 m tall. For a long time it could not be found, and was thought to be extinct, but in 1957, after almost fifty years, three Chilean ornithologists discovered James's Flamingo at a height of 4 500 m in the Bolivian Andes. The nesting habits of this flamingo are similar to those of related species.

Black-necked Swan

BLACK-NECKED SWAN

Cygnus melancoryphus

The Black-necked Swan is found in lakes and lagoons in low-lying pampas ranging from southern Brazil and southern Bolivia southwards to Patagonia. It also occurs on the Falkland Islands. In winter, it leaves the southernmost territories for northern regions. It is a heavily built bird, reaching a length of 80 cm and a weight of 4 — 6 kg. Its wingspan is about 2 m. The nesting season lasts from July to November in South America and from August to Oc-

Coscoroba Swan

tober on the Falkland Islands. The pairs defend large territories, and build bulky nests in shallow water, far from the shore. The nests are made of plant material and are hidden in dense water vegetation. The female lays 3 — 6 glossy yellowish eggs and carefully surrounds them with down. She incubates the clutch alone for 34 — 36 days, sometimes adding more down to protect the eggs from cold. The young leave the nest as soon as they have dried, and look for food under the watchful eye of their parents. Like the adults, they feed on tender leaves. Families gather into flocks of 20 — 40 in places where there is plenty of water. In dry summers, the rivers near Buenos Aires teem with up to a thousand black-necked swans.

COSCOROBA SWAN

Coscoroba coscoroba

The Coscoroba Swan is distributed from the Falkland Islands and Tierra del Fuego northwards to the Brazilian state of Rio Grande do Sul and to central Chile. It weighs about 4.5 kg. The sexes are similar in appearance, but the male has an orange-yellow iris, while the female a dark brown one. These swans frequent lakes in the pampas. In winter they leave the southern tip of the continent in flocks of about 50, and fly to southern Brazil and Bolivia, returning in the breeding season to their nesting grounds in the south, where the pairs occupy their territories. In Patagonia nesting takes place from October to December, while in Argentina these swans nest in late June. The nest is built in shallow water near the shore, or sometimes on an islet. It is 1 m across and is made of aquatic plants and twigs. The nesting cup is at first lined with fine grass, and later with down. The female lays 6 — 8 greyish-white eggs and sits on them for 27 — 35 days, the time taken to incubate them depending on weather conditions, especially temperature. Meanwhile the male stays near the nest and guards his mate. When the female leaves the nest to feed, she carefully covers the eggs with a thick layer of down so that they do not get cold. The hatched young are supervised by both parents. Coscorobas feed on grass, water vegetation, seeds and berries.

Cuban Tree Duck

CUBAN TREE DUCK

Dendrocygna arborea

The Cuban Tree Duck lives on the Greater Antilles, the northern islets of the Lesser Antilles and on the Bahamas. It measures about 35 cm, has a wingspan of 95 cm and weighs 1.2 kg. Both sexes are alike in coloration. This duck lives in swamps and lagoons in areas where there are trees. The nesting season takes place from February to September, the birds making their nests in thick vegetation or in tree hollows. The female lays 8 — 14 whitish eggs and incubates them for 30 days. Tree ducks feed on seeds, berries and insects.

Red-billed Tree or Whistling Duck

Orinoco Goose

RED-BILLED TREE or WHISTLING DUCK
Dendrocygna autumnalis

The Red-billed Tree or Whistling Duck is distributed from southern Texas to northern Argentina. It is 33 cm long and has a wingspan of 90 cm. It lives on lakes, rivers and swamps, overgrown with lush vegetation. These ducks make their nests in tree hollows, the female lining the nesting cup with down. She incubates a clutch of 8 — 16 whitish eggs for 27 days. This species is active mainly at night.

ORINOCO GOOSE
Neochen jubata

The Orinoco Goose is native to the northern half of the South American continent, but it is not found along the coast. It occurs mainly in the basins of the Orinoco and Amazon rivers. Here it exists in only a few localities, but within them it is abundant. It measures about 75 cm, the male being more robust than his mate. This goose lives beside rivers and in marshes surrounded by woods. It is chiefly vegetarian, only occasionally catching aquatic insects, molluscs and worms. For most of the year, the geese live in pairs and defend their permanent territories. Nesting takes place in the dry season. The female finds a suitable hole in a tree, near to the ground. She lines the nest with down and lays 6 — 10 cream-coloured or pale green eggs, which she incubates for 28 — 30 days. The chicks immediately feed independently but are watched over by both parents. After nesting, the birds stay for some time in flocks of 5 — 20.

FLIGHTLESS STEAMER DUCK
Tachyeres pteneres

The Flightless Steamer Duck is a large bird weighing over 6 kg, the female usually being rather smaller

Flightless Steamer Duck

White-cheeked or Bahama Pintail

than the drake. This duck has short, rounded wings, and is incapable of flight, though it flaps its wings energetically to gain speed on land. Despite its stout body, it moves surprisingly quickly, though it can cover only short distances before becoming exhausted and is an easy victim for predators if it cannot reach water in time. However, it often resists small carnivores and even chases away foxes. Water is its element. It is an expert swimmer and obtained the name 'steamer duck' from its way of propelling itself along the surface using both wings and feet, rather in the manner of a paddle steamer. It can also dive to a depth of several metres. This species is found in coastal regions of South America and on islands off Chile and Argentina, from the city of Concepcíón to Cape Horn. Despite its extensive distribution, this duck is restricted in numbers.

It lives in pairs, which vigorously defend their nesting territories, the males sometimes fighting ferociously, and hitting each other with their bills and wings. The nesting period in the southern parts of the continent takes place from September to December. The nest is built in the grass, among stones and always near to water. The nesting cup is sparsely lined with pieces of stalk or seaweed. The female lays 6 — 12 cream-coloured eggs, and incubates them while the drake swims nearby and chases away intruders. A sitting duck often allows man to approach the nest, escaping only at the very last moment. The chicks hatch after 30 days and their mother takes them to the water, where they are watched over by both parents. Steamer ducks forage for food mainly in water, feeding almost entirely on molluscs, crustaceans and other sea animals which they easily crush in their stout bills.

Rosy-billed Pochard

WHITE-CHEEKED or BAHAMA PINTAIL
Anas bahamensis

The White-cheeked or Bahama Pintail lives on the Caribbean islands, along the eastern and western South American coast, and on the Galápagos Islands, existing in several subspecies. It weighs a mere 350 — 720 g. It dwells on fresh-water lakes or river inlets, where the banks are richly covered in vegetation, but it also visits mangrove thickets. The northern populations are migratory and overwinter on the equator. For most of the year, pintails live in flocks, only forming pairs in the nesting season. The nest is built in thick vegetation near water or on an overgrown islet. The female lays 8 — 12 pale brown eggs and surrounds them with down. She incubates the clutch for 25 — 26 days and then rears the young by herself. The male often has two mates. The Bahama Pintail feeds mainly on plants, but the young also eat insects, small crustaceans and molluscs.

ROSY-BILLED POCHARD
Netta peposaca

The Rosy-billed Pochard inhabits Argentina, Chile and Uruguay. The female is predominantly dark brown above and spotted below, and weighs about 1 kg. The colourful male weighs 1.2 kg. The female builds a nest in a thick tuft of swamp vegetation, and lays 10 — 14 eggs. She incubates the clutch for 25 days. Some females lay their eggs in the nests of other species, and do not incubate them themselves. The diet of this pochard consists mainly of green plants, and seeds, though the young also feed on insects and crustaceans.

SOUTHERN CRESTED SCREAMER
Chauna torquata

The Southern Crested Screamer, which reaches a length of 90 cm and a weight of 3 kg, is characterized by its sturdy feet and the two strong sharp spurs on the forward edge of each wing. The longer upper spurs are about 5 cm long, and are used when attacking an enemy. Both sexes are alike in coloration. This species in distributed in eastern Bolivia, Paraguay and from southern Brazil to central Argentina. It is not confined to marshland and river banks, but also occurs in open pampas, often near herds of cattle or sheep. Outside the nesting season, the birds live in large flocks of up to 100. These flocks often sing in concert on river banks in the evenings and their screaming calls can be heard over distances of several kilometres. Screamers prefer to move on the ground, but when in danger, they take off rather heavily before flying to considerable heights where they even glide in the air. They are good swimmers although their feet are not webbed. In the nesting season screamers form pairs, and build a large nest of twigs and swamp plants on the shore or in shallow water. The female lays 2 — 5 yellow-white eggs, and incubates them alternately with the male for 6 weeks. As soon as they are dry the chicks leave the nest, and the brood is reared by both parents. These birds feed on small invertebrates, and on berries and green plants.

Southern Crested Screamer

HORNED SCREAMER
Anhima cornuta

The Horned Screamer is about 85 cm long. It has an unusual frontal projection up to 15 cm long and 1 cm wide, which curves forward over its bill. Its other peculiarity is the possession of two spurs at the bend of each wing, the upper spurs being 4 cm long

Horned Screamer

Everglade Kite

and sharply pointed. The Horned Screamer ranges over the tropical northern part of South America, from Colombia and Venezuela to central Brazil. When they are not nesting, the birds form flocks of up to 100, which live in sparse woodland, often on the edges of rivers and in swamps. They are ground feeders, pecking up insects, molluscs, worms and spiders, and occasionally catching small iguanas and geckos. They also eat fallen berries.
In the nesting season, screamers form pairs. They often produce strong, bubbling calls, as they vigorously shake their heads. The loud, harsh notes can be heard even far away from the trees where they roost. Screamers nest on the ground on river banks or in swamps, the nest being made of twigs, rushes and reeds. The female lays 2 white eggs and incubates them with the male for 6 weeks. The chicks leave the nest the day after hatching, and follow their parents to a place of safety where there is also sufficient food.

Long-winged Harrier

EVERGLADE KITE
Rostrhamus sociabilis

The Everglade Kite lives in freshwater marshland in lowlands ranging from southern Florida and eastern Mexico to the pampas of Uruguay and Argentina. It also occurs on Cuba. This raptor is about 38 cm long and has a wingspan of 110 cm. It has a long, slender, hooked beak adapted for removing snails from their shells. The Everglade Kite feeds entirely on a single species of freshwater snail of the genus *Pomacea,* eating as many as 200 in a day. It hunts them in the morning and evening when they are most active, by flying slowly about 5 cm above the surface of shallow water. When it sees a snail, it grasps it in the claws of one foot and lands in a dry place to eat it. It sometimes walks in the water, submerged to its abdomen. It is quite abundant in some localities, especially in the extensive swamps of the Chaco Central region of northern Argentina. Up to 200 kites may gather in one place, and during the day, they fly over the pampas in large flocks. A single dried-up tree can serve as a resting place for up to 20 birds, and they also perch on poles along the roads.
In the courtship period, these kites perform aerial acrobatic displays. The nest is built by both partners. It is situated 1 — 5 m high in a bush surrounded by water, often among reed beds. It is about 30 cm across, and is made of twigs and stems. Sometimes only the female constructs the nest while the male keeps her company and feeds her. She lays 2 — 5 whitish eggs, densely brown-spotted. Both partners incubate the clutch for 26 days and feed the young, which leave the nest at the age of one month.

LONG-WINGED HARRIER
Circus buffoni

The Long-winged Harrier is an inhabitant of the extensive reed beds of tropical South America, from

Colombia south to the Argentinian pampas. This handsome raptor is about 45 cm long, and has a wingspan of 110 cm. The female differs from the male in having brownish coloration. The harrier preys on small birds, mammals, reptiles and frogs. It also feeds on the eggs of wading birds, and on insects.

The pair builds a nest of plant material in reed beds, or on low, flat plants. The female lays 3 — 6 bluish-white eggs, and sits on them for 30 days, while the male keeps her supplied with food. When the young hatch, the male brings food which he gives to the female, and she divides it into portions for the chicks. The mother begins to hunt when the young are 21 days old, and by the age of 40 — 50 days, the young harriers are able to fly and can leave the nest. They follow their parents for another 3 weeks, continuing to be fed while they are learning to hunt.

Black Hawk

BLACK HAWK

Buteogallus anthracinus

The Black Hawk mainly inhabits the tropical, coastal regions of South and Central America, but it is also found in savannahs and in the southern United States, near rivers. It is one of the most common raptors in Honduras. It measures 50 cm, weighs 1 kg and has a wingspan of 120 cm. Both sexes are alike in coloration. This species specializes in hunting for crabs in lagoons and on the coast. It flies above shallow water, and when it sees a crab, it dives, seizes the prey with its claws, and kills it with its beak. It also hunts frogs, small mammals and small slow-flying birds. It also occasionally catches insects, including large wasps.

The nest is built 5 — 33 m above the ground, preferably in a palm tree. The base is made of branches, and the nesting cup is lined with green leaves and tiny twigs. The pair usually uses the same nest for several years, repairing and enlarging it every year. The female lays 2 eggs, grey-white with pale and dark brown spots. She incubates the clutch for 35 days, while the male brings food for her. The male also hunts food for the young, giving the prey to his mate who feeds it to them. Young hawks leave the nest at the age of 50 days.

FISHING BUZZARD

Busarellus nigricollis

The Fishing Buzzard is distributed throughout the lowlands from Mexico to Paraguay and northern Argentina. It occurs near rivers and lakes, but also be-

Fishing Buzzard

Limpkin

side brackish lagoons. It is 45 cm long and has a wingspan of 110 cm. Both sexes are alike in coloration. This beautiful raptor has feet adapted for fishing, its long toes having very strong, sharply curved claws like hooks. It flies above the water, and swoops down to the surface to seize its prey. It sometimes dives under the water, but then it has difficulty in flying up again and has to flap its wings vigorously as it pushes itself ashore. It catches large water-bugs and molluscs as well as fishes, and preys on small iguanas and rodents on land.

The nest is built in a tall tree, usually near water. Often they choose a tall mangrove, but sometimes these birds nest in a shady tree in a coffee plantation. The nest is made of branches and lined with green leaves. The female lays 2 — 3 white eggs with buffish and russet spots. She incubates the clutch for 35 days, and the male brings food for her. He also hunts for food for the young, but the mother feeds it to them. The offspring leave the nest when they are 45 days old, though they are still not fully coloured.

Grey-necked Wood Rail

LIMPKIN

Aramus guarauna

The Limpkin is a 60 cm long marsh bird of Central America, ranging north to the south-eastern United States, and south to central Argentina and western Ecuador. It mainly inhabits wooded marshland and mangrove swamps, but it sometimes settles in dry localities. It is also found in the West Indies.

Early in the morning and after sunset, or even often at night, it makes a strident drawn-out call. During the day, it roams throughout marshes looking for large molluscs, especially those of the genus *Pomacea*. Its beak is well adapted for opening shells and extricating snails. It also feeds on insects and worms, and on small lizards.

Each pair builds a large nest of branches and stems, usually on the ground, though sometimes in a tree. The female lays 4 — 8 dull yellow eggs with fawn spots, and both partners share in incubation for 25 days. The young leave the nest soon after hatching, but the adults continue to feed them.

AMERICAN JAÇANA

Jacana spinosa

The American Jaçana is distributed from Mexico southwards to northern Argentina. It also occurs in the West Indies. It measures 25 cm, and the female is bigger than the male. Jaçanas frequent the marshy shores of rivers and lakes and marshes with pools overgrown with low water vegetation. Their long toes

enable them to run nimbly across broad, floating leaves of water plants. For this reason these birds are also called lily-trotters. They feed on insects, spiders, and other invertebrates, and they collect seeds and peck green plants.
Nesting takes place from June to February, and at this time the birds produce a clucking, far-carrying call. Each female usually has several mates. They build a flat nest from bits and pieces of stems and leaves, which floats on the water. The female lays 3 — 5 buff-coloured eggs marked with irregular black lines, but the clutch is incubated for 22 — 24 days by the male, and it is he who looks after the offspring. A second brood is usually reared.

Swallow-tailed Gull

GREY-NECKED WOOD RAIL
Aramides cajanea

The Grey-necked Wood Rail makes its way dexterously through the thick vegetation of wooded marshland ranging from Costa Rica to Bolivia. This bird is about 35 cm long, and has a thin, compressed body, stout legs, and long toes which prevent it from sinking into the mud. It is also an excellent swimmer. Insects, spiders, molluscs, and tiny fishes comprise most of its diet, but it also collects seeds and pecks green plants.
Pairs nest in dense vegetation, the nest being relatively small and bowl-shaped, and made of both dry and green plants. The female lays 4 spotted eggs and incubates them for 19 days, her partner sometimes relieving her. The day after hatching, the young disperse in the reed beds, and their parents bring them food. In bad weather and when they are resting, the chicks shelter under their parents' wings. Two broods are usually reared in one year. The Grey-necked Wood Rail is active in the evenings and at night, hiding during the day among the undergrowth.

SWALLOW-TAILED GULL
Creagrus furcatus

The Swallow-tailed Gull inhabits the rocky cliffs of the Galápagos Islands. It is about 33 cm long. This bird builds a nest lined with small pebbles, and nests every nine months. The female lays a single egg, greenish to bluish-white, with brown and purple spots and stains. The partners share in incubation for 34 days, and both look after the young gull. It begins to fly at the age of 7 — 8 weeks, but the parents continue feeding it for a few weeks longer. This gull is active at night, flying in flocks or singly above the sea and feeding on crustaceans and tiny fishes which it hunts by swooping down in a spiral movement. It is capable of sustained flight, and often travels as far as 500 km out to sea in its search for food. After nesting, these gulls wander along the South American coast, from Ecuador to southern Peru.

American Jaçana

Black Skimmer

BLACK SKIMMER
Rhynchops niger

The Black Skimmer inhabits a broad belt along the Atlantic coast, from New Jersey to Buenos Aires, and along the Pacific coast, from Mexico to Chile. It is also found on the shores of Lake Titicaca in the Andes. It frequents both fresh and salty water, especially lakes situated in the coastal regions. It is about 50 cm long, and its winter plumage is adorned by a white neck stripe. This bird is unusual in that the lower part of its bill is distinctly longer than the upper. It skims above the water, cutting the surface with its lower mandible and collecting small fishes, molluscs and crustaceans. Skimmers fly in flocks, feeding in the evening and during cloudless nights. In times of drought, when there is little water in the lakes and pools, they also hunt in shallow water during the day.

They form large colonies of up to 4 000 pairs in the nesting season, often in association with terns. The simple nest consists of a hollowed out spot on the bare sand lined with pieces of stalk. The female lays 2 — 5 buff-coloured, densely black-purple and grey-spotted eggs. She incubates the clutch almost entirely on her own for 23 days. The parents feed their offspring on tiny aquatic animals. In the young, both parts of the bill are the same size, the lower part beginning to grow longer only after several weeks. Young skimmers are greyish in colour, and until they are capable of flight, they hide among stones or in clumps of grass where they are not easily seen. Skimmers produce a loud, sharp call when in flight.

Inca Tern

INCA TERN
Larosterna inca

The Inca Tern is one of the most beautiful of the terns. It is about 43 cm long, and has striking, white elongated facial feathers and a crimson bill. It inhabits islands off the coast of southern Peru and Chile. It is not migratory and nesting can take place at any time of year. Large colonies, made up of several hundred pairs, settle on the steep cliffs around the coasts. The pairs build nests in holes or crevices in the rock walls, and if no suitable hollows are available, they make nests in thick vegetation. The nesting cup is lined with seaweed or grass stalks. The female lays 2 whitish eggs, covered with dark brown and grey spots and stains. Both partners carry out incubation for 24 days, and both feed the young for 5 weeks on the nest, after which time the young terns begin to fly and learn to hunt.

The Inca Tern feeds on tiny fishes and various sea invertebrates. It dives into the water to hunt, and it is not unusual for it to pursue other terns and steal their prey.

Green Kingfisher

GREEN KINGFISHER
Chloroceryle americana

The Green Kingfisher, which reaches a length of about 20 cm, is distributed from Mexico to Panama. It occurs in small numbers further north in south-eastern Texas and Arizona. It frequents rivers, lakes and shallow reservoirs with high sand or clay banks, in which it digs burrows. The corridor leading to the nesting chamber is sometimes over 1 m long, the birds excavating it with their beaks and feet. Kingfishers do not line their nests, and the female lays 5 — 6 porcelain-white eggs on the bottom of the nesting chamber. She starts incubating as soon as the first egg is laid, and the young hatch one at a time, after 22 days of incubation. The male relieves her at irregular intervals. The young hatch naked and blind, and their parents feed them on fish and insects until at the age of 25 days, the young kingfishers are already covered with feathers and leave the nest. The adults then teach them to hunt, while feeding them for another 2 weeks. Kingfishers hunt mainly in water. They fly above the surface, stopping in the air by fluttering their wings. When they see a fish, they plummet into the water with folded wings, and seize the prey. Insects are caught in flight.

SPECTACLED CAIMAN
Caiman crocodilus

The Spectacled Caiman is the most abundant species of caiman. It occurs chiefly in the basins of the Orinoco and Amazon, but it also reaches Central America. Its average length is 2 m, though it can reach 3 m. The female digs out a depression in soft soil on a bank, and usually lays 30 — 40 hard-shelled eggs there. She stays near the clutch during the period of incubation and guards the place, which is masked with leaves and twigs. The young hatch after 50 — 80 days.

CUBAN CROCODILE
Crocodylus rhombifer

The Cuban Crocodile is restricted to Cuba, being found especially in the swampy Zapata region. It measures up to 4 m in length. Females mature at 8 — 10 years and lay 20 — 100 eggs in depressions

Cuban Crocodile

Spectacled Caiman

American Crocodile

dug out on the shore. The young hatch after 9 — 10 weeks.

Young crocodiles feed on insects, tiny fishes and tadpoles.

AMERICAN CROCODILE
Crocodylus acutus

The American Crocodile is one of the largest of the crocodiles, some specimens reaching a length of 7.5 m. Its home is in Central America and the north-west of South America. It also lives in Florida, Cuba, Jamaica and Hispaniola. It often swims far out to sea, reaching distant islands. It is still common in some areas although it is hunted for its valuable skin. It lives in small groups in coastal lagoons and river deltas. Small crocodiles feed on fishes, but larger ones catch mammals, such as peccaries. Although they rarely attack man, the local Indians fear them.
The female lays several tough-shelled eggs about the size of goose eggs, in holes dug on the shore. The young hatch after 10 days, and immediately take shelter in thick vegetation in shallow water though many of them are caught by carnivores, raptors, wading birds and large snakes. The eggs are often taken by Indians who regard them as a delicacy. The hunters follow a female crocodile, mark the place where she laid her eggs, and later dig them out.

LOGGERHEAD TURTLE
Caretta caretta

The Loggerhead Turtle is a sea turtle, weighing up to 150 kg and having a carapace 1 m long. It is relatively abundant in the warm parts of the Atlantic, Pacific and Indian Oceans. The female lays her eggs mainly on the sandy shores of Central America. They always use the same nesting places and hundreds of turtles arrive at them at the same time. They crawl ashore at night, dig out shallow depressions in the sand, and strengthen the walls with a secretion from the cloaca. They lay 100 — 150 eggs which look like ping-pong balls, cover them with sand, and return to the sea. The young hatch after 50 — 60 days and hurry to the safety of the water.

RIDLEY
Lepidochelys olivacea

The ridley *Lepidochelys olivacea* is a marine turtle, which lives in the Atlantic, Pacific and Indian

Loggerhead Turtle

Ridley Lepidochelys olivacea

Arrau

Oceans. Its carapace is up to 75 cm long. It is common along the coast of Brazil and in the Gulf of Mexico, but it was not until 1957 that its principal egg-laying ground was discovered. Subsequently other places were found. It is interesting that ten years before a Mexican engineer called Herrera chanced to visit the place. In June 1947 he was holidaying near Tampico in the state of Vera Cruz. One day he saw a huge number of turtles approaching the shore. As he watched many of them crawled up the beach ashore and laid eggs. Herrera was an enthusiastic amateur film-maker, so he filmed the event, but did nothing further about his discovery. Many years later, he showed his film to a group of people including an expert on reptiles who was astonished to see so many turtles laying eggs in daytime. When the next time for egg-laying came round, scientists were waiting in the appropriate place and were surprised to find that they were at the largest egg-laying ground of this species of ridley, and that it did indeed lay its eggs during the day.

Between April and June, up to 40 000 ridleys come ashore in this particular place. The coast here abounds in coyotes waiting for the turtles to leave, so that they can dig out their eggs. A ridley weighing about 40 kg usually lays 100 or so eggs some 50 m above the highest level reached by the tide. The young hatch after 58 — 62 days, and immediately hurry to the sea. Ridleys feed on crustaceans, molluscs and sea vegetation.

ARRAU
Podocnemis expansa

The Arrau is one of the largest freshwater terrapins. It is found in tropical South America, particularly in the basins of the Orinoco and Amazon. Its carapace is up to 1 m long, and it can weigh up to 45 kg. It has always been most commonly found in the Orinoco basin. In the period from January to April, populations living in its tributaries journey to several islands in the middle section of the river. The arraus swim, usually upstream, from localities over 160 km away. The males are the first to reach the islands, situated near the Colombian border, and a few days later they are joined by the females. The river teems with them, and the islands are loud with noise caused by carapaces hitting against one another. In the past, thousands of arraus crowded in the restricted space, with more coming and pushing up their heads like periscopes. Indians from far and wide used to arrive in canoes to collect both eggs and small arraus. Reports indicate that during each laying period up to 50 million eggs were collected. This sounds an incredible number, but these eggs were one of the most important sources of food for the Indian tribes of the Orinoco region. The number of arraus has progressively declined, and they are now faced with extinction.

In the egg-laying season, 20 — 30 females creep ashore and cautiously examine the surroundings. A few days later, more females come to join them. Each female digs a hole 1 m wide and 60 cm deep in the sandy beach using first her front feet and then the hind ones. She excavates a smaller depression inside the hole, strengthens it with a liquid secretion from her cloaca, and lays 50 — 150 eggs. Then she buries the clutch and returns to the water. The young hatch out after 45 days, usually crawling out at night. On the shore, they are hunted by raptors, and in the water they are preyed upon by crocodiles, which dispose of many young arraus as they make their way to the distant tributaries of the Orinoco.

MATAMATA
Chelus fimbriatus

The Matamata inhabits shallow waters in northern and central parts of South America. This turtle is

Matamata

Marine Iguana

over 50 cm long, and its body is covered in warts and bumps. It has an elongated snout ending in a tiny beak and fringed outgrowths of skin on its throat. These are thought to act as bait, attracting fishes to the Matamata as it lies motionless on the bottom of a river or pool. Large specimens catch small water birds and other vertebrates while the young feed on insects, larvae and tadpoles. Matamatas confined to shallow water cannot swim, while those frequenting localities susceptible to floods, are able to do so. They are good divers, remaining submerged for up to 50 minutes.

In the breeding season the female digs a depression in the mud of the river bank and lays 50 or so eggs in tough shells. She buries the clutch and then returns to the water. The young hatch after two months and immediately crawl to the nearest shallow pool.

MARINE IGUANA
Amblyrhynchus cristatus

The Marine Iguana is restricted to the Galápagos Islands. This curious-looking, slow lizard reaches a length of 1.2 m and a weight of 12 kg. Its coloration varies with age and subspecies. It feeds exclusively on seaweed. At low tide these iguanas dive to the bottom and chew tufts of seaweed. They swim with a side to side movement of the body, using the tail as propeller, and with the feet folded close to the body. Between February and March, the females lay 2 eggs in sandy holes. The young hatch between mid-May and June, usually at night.

ANACONDA
Eunectes murinus

The Anaconda is the largest of the snakes. It reaches a weight of over 150 kg and a length of 9 m. It occurs mainly in the Amazon basin, living near rivers and spending most of the day submerged in the water, waiting for animals to come to drink. It attacks the unsuspecting prey, biting it, dragging it into the

Anaconda

water and coiling its massive body around it. It catches capybaras, young tapirs, and other mammals. Like other snakes of the boa family, the Anaconda bears live young. A large female can produce as many as 70 young about 90 cm long.

Mexican Axolotl

MEXICAN AXOLOTL
Ambystoma mexicanum

The Mexican Axolotl is an interesting amphibian, distributed in the waters of Mexico and Central America, being quite abundant in some places. In the wild it usually occurs only in its larval stage, remaining in the water throughout its life and breathing through gills. The larvae are 25 — 30 cm long, and in water with a low oxygen content they often swim to the surface to breathe. This species becomes sexually mature in the larval state, a phenomenon known as neoteny. When they are about one year old, the females lay clumps of eggs on water plants. There may be as many as 500 eggs in gelatinous coatings. After 2 — 3 weeks, larvae about 10 mm long hatch out. At first they stay almost motionless on the bottom, but then they begin to hunt for small crustaceans. Later they feed on worms and insects, and weak, slow-growing axolotls fall prey to the larger ones. The larvae grow their front limbs first and then develop the hind ones. Axolotls sometimes undergo metamorphosis, the gills disappearing and adults resembling ordinary salamanders developing, which breathe with lungs. Metamorphosis occurs when the water inhabited by axolotls begins to dry up.
Both larvae and adult specimens move very slowly. They spend most of the day lying among stones or plants, waiting for prey to come near. Albino axolotls are frequently encountered, and animals in captivity in particular often lack pigment.

SURINAM TOAD
Pipa pipa

The Surinam Toad inhabits pools and small lakes in Guyana and tropical Brazil, spending all its life in them. It has large webs on its hind feet which enable it to move easily in the water, but it is helpless on land. It lives in the mud on the bottom of the pool, blending perfectly with its background and escaping the attention of predators. Here it forages for worms, insect larvae, molluscs and tiny crustaceans, digging them out with its front feet which are equipped with sensitive claws, and swallowing its prey whole since it lacks both tongue and teeth.
In the mating season, the male produces metallic sounds to attract the female. She is slightly more heavily built than the male, and reaches a length of 25 cm. The male holds her from above with his front limbs as she begins to lay her eggs. She lays about 50 of them and as she does so, the male spreads them on the skin of her back. He then swims away and takes no further care of his family. The female remains motionless for several hours, while her skin grows to form a special chamber around each egg. The pro-

Surinam Toad

Slender-fingered Bladder Frog

tected embryos develop until they have undergone metamorphosis, feeding on proteins secreted by her skin. Finally the young toads lift their lids of skin and swim out to the surface of the water.

SLENDER-FINGERED BLADDER FROG
Leptodactylus pentadactylus

The Slender-fingered Bladder Frog reaches a length of 21.5 cm. In the tropics of South America, it lives near water or in swamps, and feeds on insects, worms and spiders. In the evening these frogs produce sharp whistling sounds, and when in danger, they make loud screeching cries. The male uses his feet to whip up the jelly around his mate's eggs to produce a floating nest on which the tadpoles subsequently feed, and in which they shelter.

HORNED FROG
Ceratophrys cornuta

The Horned Frog has a triangular flap of skin over each eye. It is a relatively large frog about 20 cm long, and is one of the more colourful species. The male differs from the female in having a large orange stripe on his back while the female's stripe is green. This frog is native to northern Brazil, Venezuela and Guyana. It lives chiefly on the ground, near water, burying itself in leaves or soft humid soil, with only its eyes above the surface. It waits without moving, for small frogs, lizards, snakes, rodents, spiders and insects to approach, and swallows them whole. This frog is very aggressive, often resisting larger animals and chasing them away. Its colours also help to deter its enemies. The tadpoles are highly voracious, feeding on insects, worms, and even tiny fishes. Several other species of beautifully coloured horned frogs are resident in the South American tropics.

SOUTH AMERICAN LUNGFISH
Lepidosiren paradoxa

The South American Lungfish is the denizen of the Amazon and its tributaries, particularly the Rio Negro, and it is also found in the Rio Paraguay and in pools in the region of Gran Chaco. The male reaches a length of 1 m, while the female is slightly larger. Lungfishes live where the water is very muddy, and where the water level drops considerably in times of

Horned Frog

South American Lungfish

drought. As the level falls they dig burrows in the mud and remain dormant there until the level rises again. They are able to breathe oxygen from the air trapped in the burrow, absorbing it through air bladders, the surfaces of which are richly supplied with blood vessels. In fact the air bladders act as primitive lungs.

As soon as the rainy season begins, lungfishes leave their shelters to spawn. The female digs a vertical tunnel, usually about 50 cm long, but sometimes as much as 1 m in length, and 15 cm wide. She lays her eggs in a spacious chamber formed at the end of the tunnel, and the male protects them. He develops gill-like growths on his pelvic fins, which give off oxygen to the water around the spawn, for the water in the tunnel tends to become stale. His saliva coagulates fine particles of mud, so cleaning the water. After a few days, the larvae emerge, breathing by means of external gills, and soon feeding on tiny worms and insect larvae. Adult lungfishes feed on large water gastropods, especially on the species *Ampullaria gigas.* They take food only in the rainy season, from April to September and at this time they produce fat stores on which they feed during their inactive period. Lungfishes are much prized as food by the natives since they make excellent eating, particularly when they contain a store of fat. The natives dig them out of the chambers in the hardened mud, locating them by the ventilating shafts leading to the surface.

RED PIRANHA

Serrasalmus nattereri

The Red Piranha is one of the most feared fishes of the South American rivers, and its predatory behaviour is proverbial. Although piranhas reach a length of only 35 cm, they attack in packs, preying on every living organism they encounter including man, and being especially attracted by the smell of blood. Piranhas are fast and adroit swimmers, ferociously attacking even large animals, such as horses or cows. Shoals of hundreds of these rapacious fishes quickly bite out pieces of skin and flesh with their sharp, forward-pointing teeth, so that soon, only neat skeletons remain. In places where piranhas are plentiful, it is dangerous even to put a finger into the water from a boat. When cattle breeders want to

Red Piranha

cross a stream with their herds, they kill a weak cow or horse and throw it into the water below the fording place. The piranhas swarm around the prey, leaving the cattle to cross the river in safety. Some South American Indian tribes carry out ritual burials, sinking their dead in baskets into water rich in piranhas, and then taking out the skeletons to bury them.

Piranhas often spawn near banks among water-plants and roots. The female produces several hundred eggs, and the fry emerge after 2 — 3 days. Small piranhas feed on insect larvae, tadpoles and other small animals.

Several species of piranhas, all equally rapacious, live in the Amazon and its tributaries, as well as in other South American rivers.

SPRAYING CHARACIN

Copeina arnoldi

The Spraying Characin is widespread in bays and coves of the lower Amazon and the Rio Para, and in small tributaries overgrown with water vegetation.

Spraying Characin

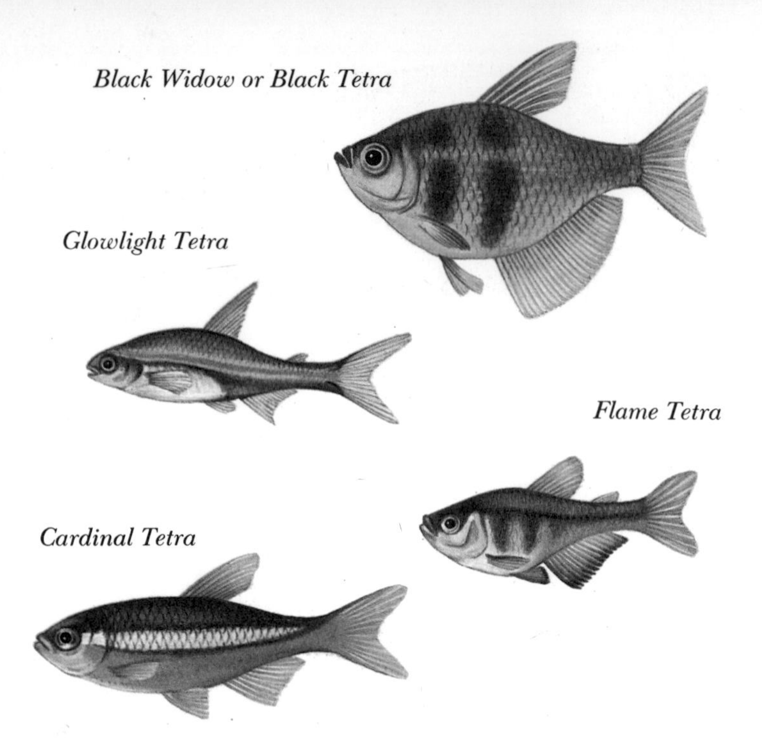
Black Widow or Black Tetra

Glowlight Tetra

Flame Tetra

Cardinal Tetra

The male reaches a length of 8 cm, but the female attains only 6 cm. This fish has an interesting mode of spawning. The male and female swim quietly near the surface, beneath the overhanging leaves of swamp plants or trees. Suddenly, both partners speed up, leap out of the water, and attach themselves to the underside of a leaf. Within a few seconds, the female lays a small number of sticky eggs and the male fertilizes them. The fishes immediately return to the water, but the spawn remains on the leaf. The partners rest for a little while and then they repeat the process, until about 50 — 100 eggs are safely stuck to the leaves. After spawning, the male takes care of the eggs, spraying them with water with his long caudal fins, to prevent them from drying up. By remaining on the leaves, the eggs are protected from aquatic predators. After 36 hours, the fry hatch and fall into the water, where they quickly hide among the aquatic plants.
The Spraying Characin feeds on tiny creatures such as crustaceans, insect larvae and worms.

BLACK WIDOW or BLACK TETRA
Gymnocorymbus ternetzi

The Black Widow or Black Tetra is native to the Rio Paraguay in the Mato Grosso region and to the Rio Negro. It is about 5.5 cm long, the female being larger and lighter-coloured than the male. Young fishes are black, while older ones are greyish, with rich black dorsal and anal fins. The female lays her eggs on water plants growing in shallow water along the bank, in a place where the water is clean. After 24 — 36 hours, the fry hatch and remain hanging on the plants for the first few days. They break loose after 5 days, and begin to swim and hunt for tiny animals, particularly crustaceans. Adult tetras feed on worms and insect larvae.

GLOWLIGHT TETRA
Hemigrammus erythrozonus

The Glowlight Tetra is a tiny fish, only 4.5 cm long, which inhabits the waters of Guyana, where it lives in shoals in shallow water along the banks of rivers and pools where the water is clear. Before spawning, tetras swim into a thick tangle of fine-leaved water plants. Here both partners rotate in the water with a screwdriver-like motion, and the female sheds her eggs when her abdomen is uppermost. The eggs become caught in the vegetation and the fry hatch after 24 hours. The tiny tetras stay in the shelter of the water plants for 3 — 4 days, feeding on their yolk sacs, before swimming out to begin foraging for food. The fry are at first a yellowish colour, which blends with the colour of the leaves of the water plants. Their diet comprises tiny crustaceans, insect larvae and worms.

FLAME TETRA
Hyphessobrycon flammeus

The Flame Tetra or 'Red Tetra from Rio' occurs in the rivers of the Brazilian state of Rio de Janeiro. It lives near the bottom, in shallow waters overgrown

Electric Eel

with vegetation. It is about 4 cm long, and is usually found in small shoals. Spawning takes place in dense vegetation, the female usually shedding her eggs on the plants, but sometimes on the river bottom. The fry emerge after 24 — 36 hours and remain on the leaves, feeding on their yolk sacs. After 5 days, the young fishes begin to swim, and become mature at the age of 4 months. The Flame Tetra feeds on tiny worms, crustaceans, water insects and their larvae, and on soft plant material. In captivity, this is a popular aquarium fish, undemanding and easy to breed.

CARDINAL TETRA
Cheirodon axelrodi

The Cardinal Tetra is a popular aquarium fish. The female reaches a length of 5 cm, but the male only 3 cm. This species was captured for the first time in 1936, in a forest pool near the upper Rio Negro. It is confined to still, unpolluted waters. It was 'discovered' by American aquarists in 1955, and shortly afterwards reached Europe, for its beautiful, neon-like red colour delighted all aquarists.
In the spawning period, the female lays 400 — 600 eggs, usually in the evening, several pairs normally spawning together. The eggs remain in thick vegetation, hidden from enemies, but they sometimes get eaten by their own parents. The fry hatch after 24 hours and shelter among the plants, venturing out after 5 days to begin to hunt for tiny crustaceans. After 5 weeks, young cardinal tetras gain their adult coloration. They are gregarious, always living in large shoals.

ELECTRIC EEL
Electrophorus electricus

The Electric Eel is one of the most interesting, but also one of the most feared of fishes. It can produce a powerful shock, by means of the electric organ which occupies four-fifths of its body. Large specimens may reach a length of 2.9 m, and these can produce a current of about 0.5 amp. at 600 volts. This easily kills small animals, and paralyzes larger ones. The Electric Eel inhabits slow or still waters in the Amazon and Orinoco basins.

Green, Brown or Cuban Rivulus

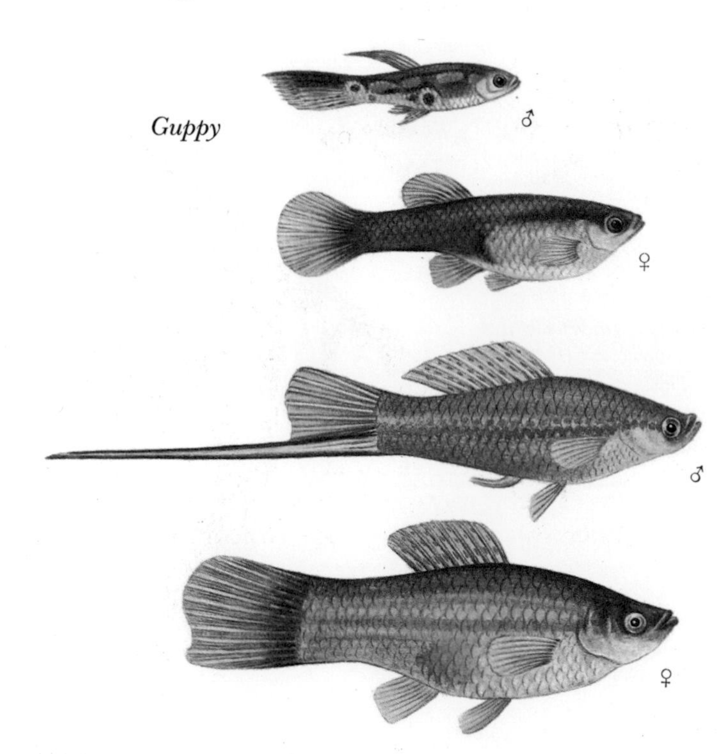

Guppy

Swordtail

GREEN, BROWN or CUBAN RIVULUS
Rivulus cylindraceus

The Green, Brown or Cuban Rivulus is a colourful fish reaching 6 cm in length. During the spawning season the male is covered with striking bright red dots. Young males and adult females have an irregular, dark, pale-edged stain at the base of the tail. The rivulus lives in sluggish waters in Cuban mountain areas, where it swims near the surface in places overgrown with vegetation. The female lays only 50 — 60 eggs at a time, but spawns several times each year. These fishes eat small insects which fall on to the surface of the water, and they often leap right out of the water to catch low-flying insects.

Jack Dempsey

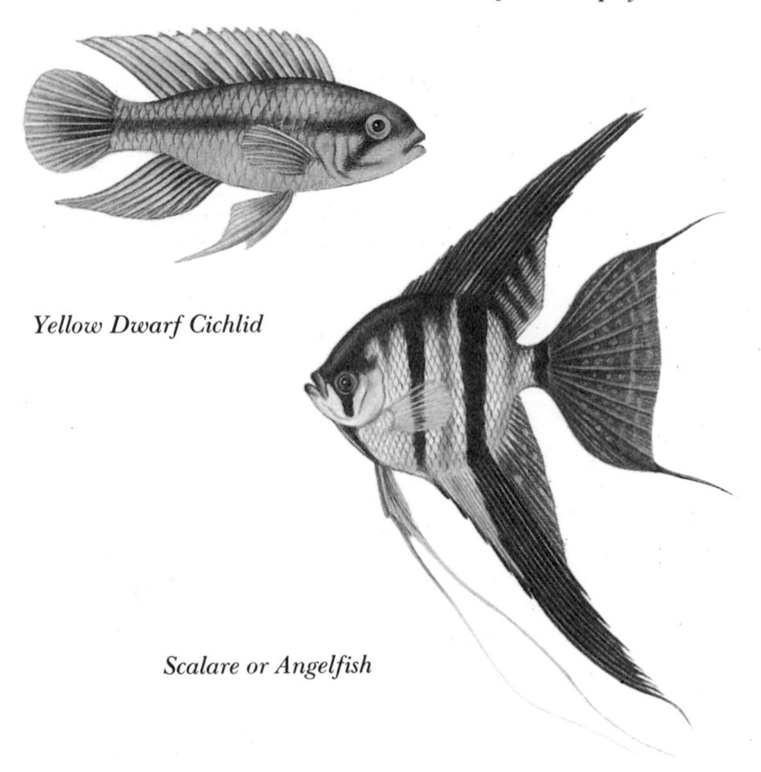

Yellow Dwarf Cichlid

Scalare or Angelfish

GUPPY
Lebistes reticulatus

The Guppy is a popular aquarium fish. It occurs in the wild in Venezuela, Guyana, Brazil and the islands of Trinidad and Barbados. The males are variable in coloration, length and shape of fins. This quality has been exploited by aquarists who have cultivated various forms with beautiful colours and shapes. The Guppy is a tiny fish, reaching a maximum size of 4 cm. The female bears 30 — 50 live young, and may produce several broods each season. The young fishes immediately scatter in the dense vegetation, hiding from predators and even from their own parents, although cannibalism in guppies is rare. The young become mature at the age of 8 months, and live for about 4 years. They feed on crustaceans, worms and insect larvae.

SWORDTAIL
Xiphophorus helleri

The Swordtail is another species of fish frequently found in aquaria. The male is about 8 cm long and the lower part of his caudal fin is elongated and sword-shaped. He has a special mating organ, modified from the anal fin, called a gonopodium. The female is 12 cm long, has a rounded caudal fin, and is stouter. Many colourful forms of this species have been bred in captivity. It is found in the wild in clear, unpolluted waters of Mexico and Guatemala. It lives near river banks or the banks of quiet pools, which are densely overgrown with vegetation. In the tangle of plants, the female bears 60 or more live young, about 8 mm long at birth. At first they feed on tiny crustaceans, adding insect larvae, worms, algae, or particles of green water plants to their diet as they grow older.

JACK DEMPSEY
Cichlasoma biocellatum

The Jack Dempsey inhabits the waters of the Amazon region. It is found near banks overgrown with water plants, in localities where the bottom is sandy and covered in large stones. The female deposits 300 — 400 eggs on clean, flat stones, and both parents care for the eggs and guard them. Before spawning these fishes dig depressions it the sand, and even uproot plants to make holes. These are used to shelter the fry, which are carried to them by the parents immediately after hatching. The fry are protected for several weeks until they are able to fend for themselves. This species is omnivorous, feeding on crustaceans, worms, insect larvae, algae and pieces of plant material. This fish reaches a length of 20 cm. The female is less colourful than the male, and in the spawning season, the male displays his shiny, deep blue-green coloration.

YELLOW DWARF CICHLID
Apistogramma reitzigi

The Yellow Dwarf Cichlid lives in unpolluted rivers and pools in the Rio Paraguay basin. The male reaches 9 cm in length, and the female 4 cm. This tiny fish dwells in localities where the bottom is covered in stones, which are necessary at spawning time. The female lays 50 — 70 red-coloured, relatively large eggs, and stays near them for 3 — 5 days waiting for the fry to hatch. When the young cichlids begin to hatch, she catches them in her mouth and carries them to holes previously dug out in a sandy area, sometimes transporting them to another locality. For 4 — 5 days the young are sustained by their yolk sacs, but then they venture out to catch minute crustaceans. Adult cichlids feed on plant material as well as on crustaceans.

Blue Discus

SCALARE or ANGELFISH
Pterophyllum scalare

The Scalare or Angelfish is another well-known aquarium fish. It is a cichlid which even in small water tanks, reaches the same size as it would in its native land. The body is markedly flattened, being much deeper than it is long. Most fishes reach 26 cm in depth and 15 cm in length. The ventral fin is elongated, becoming almost thread-like. The sexes are similar in size and coloration, but in the spawning season, the female develops a tubular ovipositor. The Scalare inhabits the middle Amazon and its tributaries, frequenting places near banks overgrown with vegetation, where it swims skilfully through the maze of plants. It often occurs in large shoals, young scalares in particular, being highly sociable. Before spawning, the partners seek broadleaved water plants with large, tough leaves, which they clean thoroughly with their mouths to remove algae and sediment. The female, accompanied by the male, attaches her eggs to the leaves. The fry hatch after a few days and fall to the bottom, fastening themselves to plants, stones, roots or various objects with a special gluey secretion produced by glands in their heads. The tiny scalares make continual darting movements with their tails to change the water round them and ensure a steady supply of oxygen. They also breathe through their tails for a few days before their gills are formed, the lower fin fold in the tail containing an auxiliary respiratory organ. Young scalares feed on crustaceans, while older ones also eat insect larvae and worms.

BLUE DISCUS
Symphysodon aequifasciata haraldi

The Blue Discus is one of the most beautiful South American fishes. It reaches a length and depth of 14 cm, its body being vertically flattened. Its coloration is variable, depending on the degree of excitement or the mood of the fish. It lives in the rivers of the Amazon region, mainly in places overgrown with vegetation. The female lays her eggs on broadleaved water plants or on flat stones, previously carefully cleaned. The eggs are guarded by both partners for 3 days until they hatch. The fry remain for another three days on the plants, before breaking loose. The tiny fishes continue to stay near their parents, and feed during the first few days of life on a mucous

Giant water-bug Belostoma columbiae

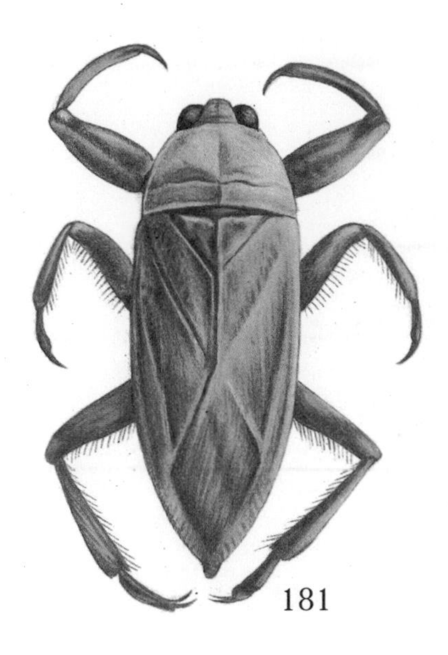

secretion from their skins. When they reach a length of about 2 cm, they begin to take other food, chiefly tiny crustaceans. Adult fishes catch worms, insect larvae and molluscs.

GIANT WATER-BUG
Belostoma columbiae

Belostoma columbiae is a veritable giant among insects. It is a water-bug, reaching a length of 12 cm! It inhabits sluggish water and river coves in the northern half of Mexico. It is an excellent swimmer, pursuing its prey underwater. It seeks tadpoles, small frogs and fishes, seizing them with its front legs which are equipped with spines. It injects poison into the body of the prey with its strong proboscis before sucking in the food. The larvae are equally rapacious, and hunt insects in the water. This water-bug also goes by the name 'toe-biter'. Although its bite is very painful, it is not dangerous to man. At night, *Belostoma columbiae* takes to the wing and can cover many kilometres in flight. Breeding takes place in water. After fertilization, the female attaches her eggs to the wing-cases of male bugs with a special adhesive substance. As she lays 150 — 175 eggs, she fastens them to as many as four males. After a few days, the nymphs hatch, leaving the males to receive further batches of eggs. The nymphs resemble adults but only gradually develop wings as they undergo a number of moults. They are able to fly after the fifth moult.

NORTH AMERICAN FORESTS

We shall now leave tropical America and travel north, through forests, across prairies and over the mountains and vast plains of North America. The landscape is more familiar to us now, and it is not unlike the countryside in Europe. Certainly it differs from it less than do the South American tropics. As was mentioned in the introduction, present-day North America was once connected to Eurasia, while the fauna and flora of South America developed separately.

When the first colonists reached North America, they were amazed at the vastness and variety of its woodland, for the primary forest covered about a third of the country. As it was less dense and more easily penetrated than tropical rainforests, exploitation of its timber was inevitable. Admittedly the native Indians felled trees to make clearings in which to form their settlements, but since they kept moving on, the clearings soon became overgrown with vegetation again. It was the new white population which affected the forests more deeply and destructively. Hundreds of thousands of hectares of woodland disappeared, so that only a quarter of it remains today, and of this only a fifth of one per cent exists in its original form, and not as a result of natural regeneration, seeding or afforestation.

The forests of the far north, in Alaska, have remained almost undamaged. Local conditions are not conducive to settlement and only about one per cent of Alaska is inhabited. The woodland here occupies the region of 40 million hectares — approximately the area of France.

The forests of North America contain a rich collection of trees. There are about 130 different species, including the tallest trees in the world.

Most common of the coniferous trees, and ranging from the south of the United States north to Alaska, are the undemanding pines, stunted forms being encountered even near the Arctic Circle. Different regions of course have their own characteristic trees. The extreme of the continent, from north-western Alaska right across to the east of Canada, is covered with vast coniferous forests. Similar forests also occur in the states of Oregon, Montana, Idaho, Wyoming and western Washington, and are even found in subtropical regions, such as Florida, South Carolina, Georgia and Alabama.

The whole of the centre, the south-east and the north-east of the United States has mixed woodlands, while predominantly deciduous woodland is found chiefly in central and eastern parts.

Firs are particularly commonly found coniferous trees, 12 species occurring in North America. Most widespread is the Balsam Fir *(Abies balsamea),* which normally reaches a height of 20 m, though stunted forms occur. It yields Canada balsam, a transparent turpentine used in optics and in the preparation of microscope slides. Another species, *Abies subalpina,* is also common, growing from Alaska to Oregon. The White Fir *(Abies concolor)* is found at heights from Colorado to southern California. It has a slender trunk and grows up to be 30 — 50 m tall.

Spruces of all kinds grow in North America. One of the most common is the White Spruce *(Picea glauca),* which covers extensive areas of central Alaska and Canada. The Colorado Spruce *(Picea pungens),* characterized by regular growth and even distribution of its branches, is widely found in Colorado, Utah, Wyoming and New Mexico. Other spruces, especially those growing in the east of the United States, are processed to make wood pulp. These spruces reach heights up to 40 m and are centuries old.

The Eastern Hemlock *(Tsuga canadensis)* is another conifer common in the mountains of eastern Canada and the east of the United States, particularly around the Great Lakes. It grows to a height of 20 — 40 m.

Larches are typical North American conifers, but these are not evergreen trees. Best known is the Western Larch *(Larix occidentalis),* which is widespread in British Columbia and Montana. This is a giant among trees, growing to a height of 40 — 60 m, and producing valuable timber.

The colossus of North American trees is the Sierra Redwood or Giant Sequoia *(Sequoiadendron giganteum),* which grows to a height of up to 90 m, and is 8 m round at the base of the trunk. The largest specimens are believed to be 3 000 years old. The remains of the original forests of Giant Sequoias which once covered the western slopes of the Sierra Nevada in California are carefully preserved in the Yosemite National Park.

These redwoods have relatively narrow, sparse crowns, with horizontal branches. When a dead branch falls it leaves a hole up to 40 cm across in the trunk. Strangely, these huge trees have only small cones, about 6 cm long. The cones of the Coast Redwood *(Sequoia sempervirens)* are even smaller, being only 2 cm long. Coast redwoods grow in the mountains along the Pacific coast, from southern Oregon to California. They are gigantic fast-growing trees, which easily regenerate from shoots round the base of felled trees, and their beautiful shiny red durable wood is in great demand. They live for nearly a thousand years.

The Bristlecone Pine *(Pinus aristata)* is the oldest living tree. Some specimens are believed to be as much as 4 600 years old, which is approximately the age of the Pyramid of Cheops in Egypt. Strictly protected Bristlecone Pines grow at a height of 3 000 — 4 000 m in the White Mountains of California.

Many varieties of deciduous trees cover large areas of North America. There are oaks, elms, and sycamores for example, majestic trees with huge trunks and widespread crowns. The Sugar Maple *(Acer saccharum)* is found in eastern parts of Canada and the United States. Its sap is the source of maple syrup, which is boiled to obtain sugar.

Black poplars are found in some localities as are willows and various species of birch. The bark of the Canoe or Paper Birch *(Betula papyrifera)* is used by the North American Indians to make their canoes. The deciduous woodlands are at their most beautiful between May and August. Then the woods are blue with Spiderwort *(Tradescantia virginiana)* and anemones glitter in the grass in spring. The pink, rich flowers of Mountain Laurel *(Kalmia latifolia)* display their beauty from May to July, and in May and June the Stemless Lady's Slipper Orchid *(Cypripedium acaule)* blooms. In autumn, shady sites are covered with blue Eastern Fringed Gentians *(Gentiana crinita)*, and forest margins are bright with lupins.

The northern coniferous forests have a rather gloomy appearance. Tufts of moss and lichens hang from trunks and branches, and softly carpet the ground. However, the branches are alive with squirrels, cracking the seeds from pine and spruce cones, and the largest of the deer, the Moose *(Alces alces)* wanders among the spruces of the Alaskan forests. Of a total population of about 160 000 mooses in Alaska, some 500 animals are killed annually by being in collision with trains. The Alaskan forests are home to about 5 000 wolves, and the largest carnivore in the world, the Kodiak Bear *(Ursus arctos middendorffii)* lives in the Aleutians and on Kodiak Island. This animal enjoys protection in the Katmai National Park, and there are about 1000 of these bears in Alaska.

The Canadian forests offer shelter to another rare animal, the Wood Bison *(Bison bison athabascae)*, known in America as the Wood Buffalo, which was on the verge of extinction, and is now strictly protected in the Wood Buffalo National Park. This is the second largest reserve in the world, occupying the area between Lake Athabasca and Great Slave Lake. The north-western part of the park is inhabited by about 200 Wood Bison. In another part of this National Park live about 12 000 other bison. These are crossbred with the Plains Bison *(Bison bison bison)*, which was introduced to the area in the years 1925 — 1928. The Wood Bison are regularly marked, treated and vaccinated, to ensure the preservation of the species for future generations. Needless to say, many other animals enjoy protection in the park.

However, not all the forest animals have been saved from extinction. Two species which were once common in the forests of North America are today seen only in museums. Both are birds. The first is the Carolina Parakeet *(Conuropsis carolinensis)*. This bird was widespread in woodland ranging from Florida to Virginia, and west to Texas, Oklahoma, Colorado, northern Iowa and Wisconsin, occasionally visiting New York. In the nesting season the pairs formed colonies, and flocks of over 1000 roamed through the countryside when the nesting period was over. The parakeets fed on grain in the fields and on ripe fruit in the plantations, with the result that farmers were quick to exterminate them. At the turn of the century, groups of 6 — 10 could still be found, but even they did not escape, and the last living Carolina Parakeet was observed in Florida in 1920.

The Passenger Pigeon *(Ectopistes migratorius)* met the same unhappy fate. This bird once occurred in flocks of millions. It never settled permanently in one place, because the enormous flocks could not find enough food in the same locality for two years in a row. The pigeons often travelled hundreds of kilometres to find new sources of food, and occupied huge areas of oak forest, where they fed on acorns. They also collected seeds and ate green shoots. Some of the flocks contained more than a million birds and were so huge that they obscured the sun for as long as three hours as they circled the sky. In the nesting season, the pigeons occupied an area covering hundreds of square kilometres. As many as 100 nests were sometimes built in a single tree, and even strong boughs fell under the weight of birds and nestlings, providing a feast for the carnivores. At this time of year people also came, bringing wagons and even herds of pigs with them. Encampments appeared around the edges of the forests, and the men made hunting trips in which they shot hundreds of pigeons, or killed the birds with long sticks as they roosted in the trees at night. As the chicks grew, people felled the trees to collect them, and the plundering hunters left behind many damaged carcasses to be devoured by the pigs. Hunters and farmers alike stored salted pigeons in barrels for use in winter, and companies were established which employed special pigeon hunters. In 1879 one company alone bought over half a million young pigeons (squabs) and there were scores of such companies as well as thousands of individuals involved. Tens of millions of squabs were killed every year. No wonder then that the gigantic flocks were soon eliminated. Passenger Pigeons were last shot in 1907, and the very last specimen died in 1914 in the Cincinnati Zoological Garden in Ohio.

In these more enlightened days, endangered species are protected in the many extensive national parks and reserves. Perhaps best known is Yellowstone National Park. This covers an area of 888 708 ha on the boundary of Wyoming, Idaho and Montana. Here the mountains are clad with conifers to a height of 2 700 m. Firs and pines predominate, but junipers are common in some localities. The deciduous woodlands contain ash trees and alders. Californian lilac of the genus *Ceanothus,* with tiny bluish or pinkish-purple flowers, grows along the tree line, and the greenery of the mountain meadows is enlivened by a multitude of flowers.

This beautiful setting provides habitats for typical large carnivores of North America such as the Grizzly Bear and the American Black Bear. These had always been hunted by the North American Indians, who ate the meat and made clothes, moccasins and beds from the skins, but the white settlers proceeded to hunt the bears to extinction. As a result of the care given to bears in the national parks, they are beginning to multiply again and they have become a particularly popular tourist attraction.

There are many other national parks and reserves covering vast forested areas of the United States, the Olympic National Park in the state of Washington being one of them. Here rare conifers are preserved, especially pines and firs, and there are a thousand other plant species. Animals of the park include 54 species of mammals, such as beavers, otters, American Black Bears, raccoons and mule deer, and there are 140 species of birds.

North America today is an area where wild fauna and flora are protected by man. Cases of deliberate damage or accidental interference are progressively becoming fewer. Indeed some species, such as the Wood Bison, are even enjoying more regular and thorough medical care than most people in the Third World.

NORTH AMERICAN OPOSSUM
Didelphis marsupialis

The North American Opossum is the only North American marsupial. It is distributed all over the eastern and central United States, occasionally occurring along the western coast and reaching Central America. It measures up to 90 cm, including the 40 cm-long prehensile tail. Though usually found in forests, it sometimes lives near human settlements. The opossum makes its den in barns or sheds, making nocturnal foraging trips in search of poultry. It kills more chickens or ducks than it can eat because it merely licks the blood, so it is relentlessly pursued by farmers. It also eats small mammals, birds' eggs, frogs, molluscs, insects and fruit. The opossum makes a nest of leaves and twigs in a sheltered place, carrying the building material in the coil of its prehensile tail. After a gestation period of 12 — 13 days, the female gives birth to 8 — 16 young, blind and naked, only 1 cm long and weighing 2 g. They crawl into their mother's pouch, attach themselves to the nipples and develop there for 3 months, after which time they leave the pouch to eat meat from their mother's prey. Half-grown infants are carried on the female's back, coiling their tails around hers. At the age of 3.5 — 4 months, opossums become independent and mature at 6 — 8 months of age.

The opossum falls prey to raptors and carnivores but has an interesting habit of feigning death when in danger. It lies down on its side, hangs its tongue out and does not move. Since most predators hunt only live prey, they lose interest in the apparently dead opossum.

This species is hunted both for its meat and its fur, over half a million opossums being caught annually.

North American Opossum

LITTLE BROWN BAT
Myotis lucifugus

The Little Brown Bat has a range of distribution covering the whole of North America except for the northernmost regions. It is about 9 cm long and spans up to 35 cm. During the day, it shelters in hollow trees, caves, lofts and uninhabited log cabins. Several thousand bats may dwell in a single cave, often sleeping in clusters, hanging on to each other, and covering the walls in dense rows. They set off to feed at dusk, hunting insects, particularly beetles and moths, flying about 5 — 10 m above the ground. In the breeding season, the females form separate colonies for themselves and their young. After a gestation period of 55 days, the female bears a single young, or sometimes twins. She hangs in a horizontal position, and the blind naked baby crawls to its mother's chest, using its claws, and clings to her fur as she flies. When the infant is a month old and has grown its fur, the female leaves it in the colony, returning regularly to suckle it. She recognizes her own offspring among the hundreds there from its voice. At the age of 6 weeks the young bats are able to fly and they learn to hunt under their mother's supervision. Little brown bats sometimes fly to places over 200 km from their permanent quarters.

About 65 other species of bat live in North America.

AMERICAN BLACK BEAR
Ursus americanus

The American Black Bear was once found almost over all the United States. It was later exterminated in many places, but nowadays it is common in Canada and in the American north. It is rare in the Mid-

west, but extends to the south-eastern states and to Mexico.
This bear has a glossy black coat and reaches about 180 cm in length, 100 cm in height at the shoulder, and has a weight of 150 kg. The female is more lightly built than the male. For most of the year this bear is solitary. It roams through its territory hunting small vertebrates on land and in the water. It mainly eats fish, but also feeds on molluscs and insects as well as on sweet forest fruits. It attracts tourists to national parks, coming to roads and clowning for sweets. Occupants are forbidden to leave their cars in such places, for the seemingly tame bears can be very dangerous.
Black bears sleep through the winter in caves or in hollows under uprooted trees. They take no food at this time and live on fat which they put on before the onset of winter. After a gestation period of 7 — 7.5 months, the female bears 2, or sometimes up to 4 young usually in January or February while she is still in her den. The cubs are born sparsely furred, blind, and weighing about 300 g. They open their eyes after 28 — 38 days. The mother holds them close to the fur of her abdomen to keep them warm, and suckles them. She takes them out for the first time when they are 55 — 70 days old, and a month later the young bears begin to forage for food. They stay with their mother for two years.

Little Brown Bat

GRIZZLY BEAR
Ursus horribilis

The Grizzly Bear was a symbol of strength and courage to the American Indians. Their hunters made necklaces from its powerful claws to mark their bravery in their struggles with this formidable carnivore. There are many stories of encounters between

American Black Bear

Grizzly Bear

Indian warriors and grizzly bears but in reality man was only ever attacked by an animal which had been wounded or cornered. If left undisturbed this bear avoids man, although in national parks it has learned how to beg for sweets from tourists. The Grizzly Bear was once found throughout the west and mid-west of the United States and Canada, as well as in north-western Mexico. It was exterminated in many places, and is now restricted in the United States to a small locality in the mountains of Montana. Larger populations live in Canada and Alaska.
The Grizzly Bear is up to 2.5 m long and weighs over 300 kg. It is solitary, forming pairs only in the mating season, from April to July. It sets out to hunt after sunset, though when the salmon are running, it fishes during the day. It searches for large vertebrates, birds and reptiles, but it also eats fruits, flowers and honey. It sometimes digs grubs out of decaying stumps or finds slugs. It spends the winter in its den, sleeping through unfavourable weather. This is not a true hibernation, since its body temperature does not fall. In January or February, after a gestation period of 7 — 8 months, the female gives birth to twins or triplets. The cubs weigh about 400 g and open their eyes after 29 — 35 days. At the age of 75 — 90 days, the young leave the den for the first time. They are very active and playful and are fond of climbing trees. They mature after 2.5 years.

KODIAK BEAR

Ursus arctos middendorffii

The Kodiak Bear is the largest of the bears, which makes it the largest carnivore. It can weigh over 1 000 kg and measure up to 3 m. It is 1.2 m tall at the shoulder. This giant inhabits Kodiak Island, off the west coast of Alaska, and also lives in the western part of the Alaskan Peninsula. Kodiak bears are solitary for most of the year, living in pairs only in the mating season, in June or July. At the end of October, the bears look for suitable sheltered places to spend the winter, and build dens in caves or in hollows under the roots of large trees. The female gives birth to 1 — 6 young in the shelter, in December or January, after a gestation period lasting 7 — 8 months. The young gain sight after 29 — 35 days. At the age of 45 days, the cubs leave the nest for the first time, and when they are 3 months old they begin to take solid food brought by their mother.

Kodiak Bear

RING-TAILED RACCOON
Procyon lotor

The Ring-tailed Raccoon inhabits almost the whole of the United States and northern Mexico. It frequents deciduous and mixed woodlands where there are lakes and rivers. This small bear-like carnivore reaches a length of 90 — 100 cm including the bushy tail. It lives in pairs or small groups composed of one male and several females. In the daytime it usually shelters in a hole in a tree or sleeps in a nest deserted by raptors, though sometimes it just settles on a bough in a treetop. It forages after sunset, climbing over trees, scurrying on the ground or poking in shallow water. It has a very good sense of smell, which helps it to locate nests containing eggs or nestlings. It catches frogs, crustaceans and fishes, and takes domestic poultry. It also eats fruit and berries. The raccoon often carefully washes its prey in water before devouring it.
In April or May, after a gestation period of 63 — 64 days, the female bears 1 — 7 young in her burrow. The young are born blind and obtain their sight after 24 days. They leave the burrow after 40 — 45 days. They are suckled for 16 — 18 weeks, but from the age of 55 days, they begin to eat pieces of prey.

Ring-tailed Raccoon

American Pine Marten

Populations resident in northerly regions usually sleep through the coldest periods, living off their stores of fat. They often wake up during spells of mild weather, and venture out. Raccoons are extensively hunted for their beautiful fur.

AMERICAN PINE MARTEN

Martes americana

The American Pine Marten lives in Canada and in the north-east of the United States. It used to be common in the states of Washington, Oregon and Montana, but was exterminated in the quest for its valuable, high-quality fur. This marten reaches a length of about 70 cm. It is solitary, mainly frequenting pine forests, but it is sometimes found in mixed woodland. During the day, it sleeps in a hollow in a tree, usually high above the ground. It leaves the nest after sunset to run nimbly among the branches. It marks out its territory with a strong-smelling secretion. It is expert at catching squirrels and birds, and searches for birds' eggs and sweet fruit.

In summer, in the breeding season, the martens seek partners, their barking voices being heard far afield on clear nights. After a gestation period of 260 — 300 days, the female gives birth to 3 — 5 young, usually in April. The young obtain their sight after 34 — 38 days. They are suckled for 6 — 7 weeks, and then feed on prey brought by the female. Young martens mature at 27 — 28 months.

Fisher Marten

FISHER MARTEN

Martes pennanti

The Fisher Marten is distributed in Canada and in the north-west and north-east of the United States. It used to be common around the Great Lakes, but it has been hunted to extinction for its beautiful fur.

The Fisher Marten is about 100 cm long and weighs 1.5 kg. It has a strong tail, and dense, fine fur. It lives in forests near lakes, rivers or brooks, but its name is misleading.

Although it is an excellent swimmer, it does not catch fish at all. It is adept at climbing and finds most of its food in trees, catching small mammals and birds, and eating fruit. It builds up stores of food to eat when it is hungry. The marten seeks its food after sunset, sleeping throughout the day in a hollow tree. Its territory is marked by a foul-smelling fluid exuded from vents under the tail.

After a gestation period of 260 — 268 days, during which the embryo ceases to develop for some time (delayed implantation), the female gives birth to 3 — 5 young. These are blind at first, and open their eyes after 34 — 38 days. The young martens soon begin to take solid food. Maturity is reached at the age of 27 — 28 months.

Wolverine

WOLVERINE
Gulo gulo

The Wolverine is a marten-like carnivore, about 1 m long. It is common in Canada and Alaska. At one time it inhabited the mountains of Montana and the region round the Great Lakes in the American north-east. The Wolverine is usually solitary and is active throughout the year. It is a good climber and swimmer. It feeds on small mammals and birds, occasionally attacking larger animals. During the day it hides in underground shelters, lined with leaves. After a gestation period of 260 — 320 days, the female bears 2 — 4 young, usually in February or March. The young are born blind and open their eyes after 28 — 35 days.

AMERICAN BADGER
Taxidea taxus

The American Badger reaches a length of 65 cm and a weight of 10 kg. Its range covers the woods of the centre and west of the United States, northern Mexico and southern and central parts of Canada. It digs out deep burrows with tunnels several metres long, and containing a nesting chamber lined with leaves and grass. Here the badger sleeps during the day, setting out to feed after sunset. Badgers usually live in pairs throughout the year, hunting together for rodents, small birds, beetles and worms. They also feed on various fruits, dig out roots, and occasionally take bird's eggs. In May or June the female bears 3 — 4 blind young, which obtain their sight after 4 — 5 weeks. The gestation period lasts 7 — 8 months (sometimes 12 — 15 months), though the embryo develops for only 60 — 70 days. The young are suckled for 8 — 10 weeks and stay with their mother for 7 — 8 months. They are mature at 1.5 years.
Badgers keep their burrows meticulously cleen, burying urine and faeces in a hole near the nest. In the extreme north, these badgers sleep through part of the winter, often waking up and walking in the deep snow before falling asleep for a further few days. At this time they live on their stores of subcutaneous fat.

NORTH AMERICAN RED FOX
Vulpes vulpes fulva

The North American Red Fox is about 1 m long, and inhabits Alaska, Canada, and a large part of the United States. It prefers thinly forested areas, and frequents bushy localities with scattered rocks, particularly near lakes. The fox marks out its territory with urine and then digs out a burrow or earth with several emergency exits, or settles in a small cave. Its den is easily recognized by the heaps of discarded remains of prey lying at the entrance. Foxes sometimes excavate several other dens within their territo-

American Badger

North American Red Fox

ry, but these are only used in times of danger. Although it is nocturnal, the fox likes to bask in the sun in front of its den. It comes out to hunt after sunset, chiefly catching small mammals, birds, insects and molluscs. It occasionally eats berries and fruit.
The courtship period takes place from January to March, two contenders sometimes fighting for one female. Partnerships often last for several years. The female lines her den with fine hair torn from her abdomen. After a gestation period of 51 — 52 days she gives birth to 3 — 8 young, usually in April. The young are born blind and gain their sight after 12 — 15 days. They are suckled for 8 — 10 days. When they begin to take solid food, their parents are kept busy, finding food for their hungry offspring, and at this time foxes cause considerable damage to both game and poultry.
As well as the normal red form, this species is found in silver and black forms, and one litter may contain all three colours. The silver variety possesses particularly beautiful fur and is bred on farms.

WOLF
Canis lupus

The Wolf is distributed throughout North America, from Alaska to northern Mexico. The largest wolves, which live in Alaska and Canada, reach a length of 1.8 m, a height at the shoulder of 1 m, and a weight of 75 kg. The Wolf inhabits vast wooded areas, or bush-covered localities in the south. In February, when the mating season begins, their typical long-drawn howling can be heard from the forests. In April or May, the female bears 5 — 7 cubs after a gestation period lasting 63 — 64 days. The young are born blind in a den dug out by the female in a sheltered place. They open their eyes after 8 — 14 days. Wolves feed on mammals as large as a moose, birds, small lizards, frogs, and fruits.

Wolf

Canadian Lynx

CANADIAN LYNX
Felis lynx canadensis

The Canadian Lynx is a robust feline carnivore, reaching a length of about 1 m and a weight of 50 kg. It is found mainly in Alaska and Canada, ranging south to the states of Washington and Wyoming in the United States. It is an alert and cautious animal, hiding in the vast forests of both lowlands and mountains. It hunts after dusk, occasionally leaving its shelter during the day to bask in the sun. It combs its territory in search of small mammals, especially snow hares. It catches birds, especially ground-feeding species, and occasionally takes snakes. The lynx lies in wait for larger mammals, jumping on its prey from a rock or a strong bough, and dragging it to the

Bobcat or Bay Lynx

ground with its weight. It even attacks young deer, or an older, weak adult. The lynx cuts the neck artery of the prey with its teeth, or breaks the neck of a smaller animal.

The lynx is solitary except in the mating season when the males seek mates. Following a gestation period of 10 weeks, the female bears 1 — 4 young in April or May. She shelters them in a spacious tree hollow or in a cave. The kittens are born blind and gain their sight after 10 — 14 days. They are suckled for 13 weeks. Young lynxes mature at the age of 21 months.

BOBCAT or BAY LYNX
Felis rufa

The Bobcat or Bay Lynx is distributed in central and southern parts of the North American continent, being found almost all over the United States, in south-western Canada and northern Mexico. It frequents forests with rocky sites, bush-covered localities and semideserts. It measures over 90 cm in length and weighs 13 kg.

WOODCHUCK
Marmota monax

The Woodchuck is the only marmot found in woodlands. It is widespread in Canada and in north-east and eastern parts of the United States. Its body reaches a length of 65 cm and it weighs up to 6 kg. The female is smaller than the male. The Woodchuck has a thick coat and feet well adapted for digging. It makes underground burrows for shelter in times of danger and at night. Woodchucks are diurnal, beginning to browse immediately after sunrise, and basking in the sun for hours. During bad weather, woodchucks may stay in their burrows for several days. In early October they carefully fill the entrance with stones, pieces of turf, and grass, and hibernate until February.

The Woodchuck feeds on grass, leaves, shoots, seeds and various fruits, being often found in plantations and gardens. Despite its keen hearing and eyesight, this cautious animal often falls prey to foxes, lynxes, wolves and large raptors.

After a gestation period of 31 — 32 days, the female bears 2 — 6 young in a burrow lined with leaves, generally in April. The young are born blind and open their eyes after 20 — 28 days. They leave the burrow when they are 40 days old and suck their mother's milk for 8 — 10 weeks.

GOLDEN-MANTLED GROUND SQUIRREL
Citellus lateralis

The Golden-mantled Ground Squirrel measures 25 — 28 cm. It is distributed in the west of the United States and in south-western Canada. It frequents coniferous forests and rocky localities, often living in colonies. It makes burrows several metres long and as deep as 1.5 m. This species feeds mainly on the seeds of coniferous trees, especially pines, and on forest fruits. After a gestation period of 28 days, the

Woodchuck

Golden-mantled Ground Squirrel

female gives birth to 3 — 8 young in an underground nest, lined with grass, leaves and feathers. The young are born blind and gain their sight after 7 — 8 days. They are suckled for 4 — 6 weeks.

Eastern Chipmunk

EASTERN CHIPMUNK
Tamias striatus

The Eastern Chipmunk is a common rodent of the woodlands of the east, midwest and north-east of the United States and south-east Canada. It is 25 cm long. The chipmunk is predominantly terrestrial, digging burrows beneath roots or among boulders, usually building several shelters within its territory. It sometimes climbs on low branches of trees or shrubs in search of nuts, seeds, berries and insects. The litter is born in a burrow up to 3 m long with a nesting chamber at the end, which is dug out by the female. She bears 3 — 5 blind young after a gestation period of 30 days. Young chipmunks open their eyes after 7 days and leave the burrow at the age of one month. Winter is spent in hibernation.
The North American territory is inhabited by many other species of chipmunk, especially those of the genus *Eutamias.*

Grey Squirrel

GREY SQUIRREL
Sciurus carolinensis

The Grey Squirrel inhabits the eastern half of the United States, where it lives in forests, parks and large gardens. It builds its nest of twigs, leaves and moss in trees or hollows. After a gestation period of 35 — 44 days, the female gives birth to 4 — 6 young in the finely lined nest. The newborn squirrels are blind, opening their eyes after 30 — 35 days.

AMERICAN FLYING SQUIRREL
Glaucomys volans

The American Flying Squirrel is a small rodent, distributed in the vast woodlands of the eastern half of the United States. It is 25 cm long, including the tail, which measures 10 cm. It has a membrane stretched along each flank between the front and hind limbs. When the squirrel is threatened from above, it jumps

American Flying Squirrel

North American Porcupine

down, extends the membrane, and glides for distances of 20 — 30 metres. On alighting, it immediately climbs the nearest tree, for it moves clumsily on the ground.
The flying squirrel has the characteristic large eyes of the nocturnal animal. During the day it hides in a hollow in a tree, and sets out after dusk to feed on insects, birds' eggs, nuts and various forest fruits. After gestation periods of approximately one month, the female bears two litters of 3 — 6 young annually. This species is preyed upon by large owls and martens.

NORTH AMERICAN PORCUPINE
Erethizon dorsatum

The North American Porcupine inhabits coastal woodlands of Alaska, Canada and almost all of the western half of the United States. This rodent reaches a length of 80 cm and a weight of up to 20 kg. It has a small head, a sturdy body and a short tail, and it has short, barbed spines in its coat.
The North American Porcupine is predominantly a tree-dweller, slowly climbing on branches in the treetops. When descending to the ground it moves in short, rapid hops. It forages after dusk, feeding on fruits, seeds, shoots and leaves, and nibbling bark, which it peels downwards from the branches. In places where it is abundant, this animal can cause considerable damage to young trees. During the day, it sleeps in a hollow in a tree or in a forked branch. After a gestation period of 210 — 217 days the female bears a single young which opens its eyes immediately after birth and already has a thin coat of fur. The infant is well developed, and is able to follow its mother after two days.

SNOWSHOE HARE
Lepus americanus

The Snowshoe Hare lives in the forests of Canada and the north of the United States, from which it

Snowshoe Hare

ranges into the states of Utah and Nevada. It measures 32 — 45 cm, and it is characterized by its seasonal change of coloration. It becomes pure white in winter, and grows long, thick hairs on its feet, which spread its weight on the snow like snowshoes. In early spring the hare moults into a brownish coloration. In the breeding season the female prepares a nest in grass or under a bush, concealing it from enemies. After a gestation period of 35 — 37 days she produces 2 — 7 young, born with open eyes and covered with fur. The mother suckles them for 14 — 21 days, but the young often begin to eat grass and leaves when they are a week old.

White-tailed or Virginian Deer

WHITE-TAILED or VIRGINIAN DEER
Odocoileus virginianus

The White-tailed or Virginian Deer is very common throughout North America, being found in southern Canada, and through most of the United States except for the extreme west, though it occurs in the states of Oregon and Washington. In the south it reaches Mexico. It is a forest-dweller, sometimes frequenting bush-covered localities. The male is up to 110 cm high at the shoulder, weighs 200 kg and has basket-shaped antlers. The female is smaller, more delicately built, and lacks antlers. This deer has a relatively long tail, measuring about 30 cm, which is white-coloured below. When in flight, herds of these deer raise their tails to show their shiny white colour. This enables the herd to stay together, especially at dusk, when the signals become particularly important.

Wapiti

In the rutting season, each male usually has a harem of 3 — 4 females, but in winter the animals form larger herds. Between May and August, the female bears a single young or twins after a gestation period of 196 — 210 days. The young deer is born with open eyes and is able to walk, though it remains lying in a sheltered place for 5 — 10 days while the mother comes to suckle it. After 10 days the beautifully spotted young deer follows its mother, who continues to suckle it for 1 — 2 months. Young female does mature at the age of 2 years, the males at 3 years. These deer feed on leaves, twigs, grass, nuts and lichens. They are hunted by man and are preyed upon by large carnivores, such as bears, mountain lions, lynxes and wolves.

WAPITI
Cervus canadensis

The Wapiti is a robust deer which reaches a height of 150 cm at the shoulder and a weight of over 300 kg. The female is slightly smaller. [Some zoologists classify the Wapiti as a Red Deer *(Cervus elaphus).*] The Wapiti once ranged from south-western Canada to New Mexico as a common resident of the forests. It was, however, extensively hunted, and was eliminated altogether in many places. The rutting season begins in late August, the males engaging in duels for the possession of herds of females. After a gestation period od 240 days, the female bears a single young or sometimes twins, usually in May.

MOOSE
Alces alces

The Moose is the largest of the deer. The biggest specimens occur in Alaska, but it also lives in Canada, and in the region of the Great Lakes and in the states of New York, Montana and Idaho in the United States. The male is 235 cm tall at the shoulder, is 275 cm long, and weighs over 500 kg. He has outstandingly massive antlers up to 160 cm wide and weighing 20 kg. These antlers bear as many as 30 tines, the shape differing from animal to animal. Mature males shed their antlers in November or December, but young ones lose theirs in January or February. Following a gestation period of 240 — 250 days, the female gives birth to a single young or twins, usually in May or June. Young mooses grow quickly and mature when they are about two and a half years old.

SWALLOW-TAILED KITE
Elanoides forficatus

The Swallow-tailed Kite is widespread throughout the American continent. It inhabits the eastern half of the United States, frequenting pine forests and swampland, and is common in Mexico up to a height

Moose

of 1 800 m. Its area of distribution also covers Central and South America, where it lives in humid woodland as far south as northern Argentina. It reaches a length of 55 cm and has a wingspan of 130 cm. This graceful bird is conspicuous on account of its long, forked tail. Several kites sometimes hunt together, catching and swallowing insects in flight, and perching afterwards on a single tree, often in groups of 10 or more. During the breeding season, kites raid the nests of smaller species, taking both eggs and nestlings. These birds also occasionally catch small reptiles. In Florida, kites often come down to the ground to collect locusts and wasps, or to catch small fishes in shallow water. Kites drink on the wing in the same way as swallows, from the surface of lakes or rivers.

At nesting time pairs build their nests in tall treetops — usually in pine trees in the United States. They use twigs gathered in flight, and the nesting cup is lined with moss, carried in beaks and claws. The finished nest is about 50 cm across and 30 cm high. Here the female lays 2, or occasionally 4 russet-spotted eggs. In Florida, nesting takes place from mid-March to mid-April. The female incubates the clutch by herself, and the young leave the nest at the age of 35 — 40 days.

Swallow-tailed Kite

Red-tailed Hawk

Cooper's Hawk

COOPER'S HAWK
Accipiter cooperii

Cooper's Hawk inhabits forest areas from southern Canada across the United States to north-western Mexico. In Arizona, this raptor lives near rivers. It is about 40 cm long and has a wingspan of 80 cm, the female being considerably larger than the male. This hawk is partly migratory, leaving its northernmost territories in October, and flying south to Costa Rica, and sometimes to Colombia. It prefers extensive woodland, where it preys mainly on thrushes, bee-eaters and small woodpeckers. The nest is built among the treetops. A day or two before laying the first egg, the female lines the nesting cup with pieces of bark, peeled from a tree with her beak. Sometimes the male also brings bark. A clutch of 4 — 5 pale blue to grey-white eggs, sometimes russet-spotted, is laid, and chicks hatch after 34 — 36 days. For the first three weeks, the family is fed by the male.

Ruffed Grouse

RED-TAILED HAWK
Buteo jamaicensis

The Red-tailed Hawk is distributed from Alaska across Canada and the United States to western Panama. It also lives on the Bahamas. This hawk is found in several subspecies, differing mainly in coloration, though most adult birds of all the subspecies have red-coloured tails. The Red-tailed Hawk is about 45 cm long, and has a wingspan of about 120 cm. It frequents localities where pine and oak woodland is interspersed with meadows and fields, and it also lives near water. In spring, the pairs build large nests in the trees, lined with stalks and bark. Some pairs build nests on rocks. The female lays 1 — 4 whitish eggs, and incubates them, her mate bringing food for her and occasionally sharing in incubation. The chicks hatch after 28 — 32 days, and their parents feed them carefully. A young hawk can swallow incredibly large pieces of food. When it is only six days old it can consume a 30 cm-long rattlesnake — minus the head, which is torn off beforehand by the female. The young raptors begin to fly when they are 45 days old.

California Quail

RUFFED GROUSE
Bonasa umbellus

The Ruffed Grouse lives in Alaska and Canada, and in the north-eastern and north-western regions of the United States. It also occurs in the region around the Great Lakes. It is a common fowl-like bird, reaching a size of 40 cm. It frequents sparse woodland in summer, and dense coniferous forests in winter. It exists in two colour forms, red and grey, each having a variety of colour shades.

The Ruffed Grouse lives in pairs. In the courtship period, the male spreads his tail, and raises it so that it stands vertically, so creating the effect of a colourful wheel. The female forms a shallow depression, usually sheltered by a tree trunk, and lines it with leaves and stalks. She lays 8 — 14 spotted eggs and incubates them for 24 days. The chicks are raised by both parents. If a hen dies, her partner has been observed to look after the offspring by himself. The young feed independently, at first only eating insects, but later pecking seeds and green plants.

CALIFORNIA QUAIL
Lophortyx californicus

The California Quail is a fowl-like bird, which reaches a length of about 24 cm. It inhabits the extreme west of the United States and south-western Canada. The female differs from the male in that her head and breast are predominantly greyish. The California Quail resides in mixed woods and in large parks where there is extensive bush cover. The female builds the nest under a thick bush, lining the shallow depression with leaves and stalks. She lays 12 — 16 cream-coloured eggs, covered with golden and brownish dots and spots, and incubates them for 21 — 23 days. At first the chicks are small, but they are very active and can feed themselves. Their mother wanders with them and protects them from their enemies. At the age of 3 weeks, the young quails begin to roost on branches at night. This bird feeds on insects, spiders, molluscs, seeds, and green plants.

WILD TURKEY
Meleagris gallopavo

The Wild Turkey is a characteristic bird of the North American continent. It lives in central, southern, north-eastern and eastern regions of the United States and in Mexico. However, its distribution is discontinuous, and it can no longer be found in many places, such as Georgia and Carolina. The Wild Turkey is about 85 cm long, the female being more lightly built and lacking the prominent skin fold on the neck and head. Turkeys inhabit open woodland, or forest clearings. Courtship begins in late February, the males gathering at special sites to display, and make their typical gobbling cries. They inflate their necks, ruffle their plumage and hang down their wings, as they try to attract the attention of the females. Sometimes the cocks engage in serious fighting, and the winner keeps 2 — 4 hens in his harem. After mating, the hens disperse to build their nests and the cocks take no further part in the proceedings.
The female digs a shallow depression in a thicket, under fallen tree trunks or beneath dense bushes, and lines it sparsely with grass and leaves. She lays 9 — 12 eggs, ochre to pale fawn in colour and with dark spots and dots. Older hens lay as many as 20 eggs, and while the clutch is incomplete the hen stays near the nest, and keeps the eggs covered with leaves. She begins incubation when all the clutch is laid, leaving the nest only briefly to feed or drink. The young hatch after 26 — 28 days, and set out in search of food the following day, accompanied by their mother. Turkeys feed on small insects and larvae, spiders, tender shoots and leaves. The hen protects her offspring and at night or in bad weather she shelters them with her wings. After 10 days, when the young begin to fly, the family starts roosting in the trees. Adult birds usually gather in groups to roost.

Wild Turkey

Wild turkeys often feed on grain in cultivated fields, and on various kinds of fruit. Adult birds sometimes catch small mammals.

AMERICAN WOODCOCK
Scolopax minor

The American Woodcock inhabits damp woodlands and wooded marshland in south-eastern Canada and in eastern regions of the United States. It reaches a length of about 24 cm and has a very short neck. In

American Woodcock

Mourning Dove

the autumn, this woodcock migrates south to spend the winter in the south-east of the United States, returning to its nesting territories in late March or early April. Woodcocks form pairs during the breeding season, and the female builds a nest in a shallow depression in the ground and lines it with leaves and moss. She lays 4 ochre-coloured eggs with pale and dark brown spots, and incubates them by herself for 22 days. The young leave the nest after they have dried, and hide in dense clumps of grass or among the roots of trees. Here the parents bring prey to them for the first few days. The young woodcocks begin to fly when they are 3 weeks old. These birds seek food on the ground, probing it with their long bills and pecking insects, larvae, worms and molluscs out of the soft soil and decaying leaves. Special cells on their bills are sensitive to touch, and enable the woodcock to detect prey under the surface.

Spotted Owl

MOURNING DOVE
Zenaidura macroura

The Mourning Dove is one of the most common birds of southern Canada, the United States, Central America and the Antilles. In autumn, it migrates from the northerly areas of its distribution to the south, but it is resident in the other regions. It measures about 26 cm, and is found in thin woodland and in bush-covered localities. This dove forms pairs in the nesting season, but seeks its food in flocks of 50 — 60 birds. It feeds on various seeds and on grain as well as on green plants, occasional insects and invertebrates. The nest, made from dry twigs and situated among bushes or low trees, is a flat, casually constructed affair. The female lays two white eggs, which are incubated by both partners for 14 — 15 days, though the female spends longer on the nest. The young are fed on a special mushy substance from the parents' crops, and later they take predigested seeds and insects. The young doves leave the nest at the age of 13 — 15 days.

SPOTTED OWL
Strix occidentalis

The Spotted Owl lives in forested areas in the south-west of the United States, mainly along the coast, and in Mexico. It measures about 32 cm and has a wingspan of 105 cm. After sunset, this owl hunts various small mammals, mainly squirrels, birds, amphibians and larger insects. The nest is made in a tree hollow or in a nest abandoned by raptors. The female lays 3 — 5 white eggs and sits on them for 28 — 30 days. During this time, the male feeds her and later he brings food for the whole family. The young are capable of flight when they are 50 days old.

Great Horned Owl

GREAT HORNED OWL
Bubo virginianus

The Great Horned Owl is the largest of the American owls. It reaches a length of 55 cm and has a wingspan of 135 cm. This owl is distributed throughout the continent, from Alaska across Canada and the United States to Peru and Brazil. It is found in a variety of habitats including woodlands, mountain areas, rocks, bushland and prairies, normally being there throughout the year. A pair seeks a nest deserted by raptors, in rocks and tree hollows. The female lays 2 — 3 white eggs and incubates them for 34 days. The young are fed on the nest for about 6 weeks, after which time they start perching on nearby trees. They are able to fly at the age of 3 months. The Great Horned Owl preys on wild rabbits, prairie dogs and ground squirrels.

HAWK-OWL
Surnia ulula

The Hawk-owl is distributed in Alaska and Canada. It attains a length of about 35 cm and a wingspan of 85 cm. It has characteristic crosswise striping on the underparts, and relatively short, pointed wings. In flight, it resembles a bird of prey. It is resident in its native land, frequenting thinly forested areas and open scrubland. It usually hunts after sunset, but in winter it comes out during the day. It perches on dead branches of tall trees or on telegraph wires

Hawk-owl

Saw-whet Owl

where there is good view, often perching bent slightly forward, and flitting its tail. It mainly hunts small rodents or, less frequently, small birds. The nest is built in a hole in a tree, or in nests deserted by birds of prey, or very occasionally, on the ground. The female lays 3 — 5 eggs, but at times when a large rodent population ensures a plentiful supply of food, as many as 10 eggs may be laid. The owl chases away animals much stronger than itself, and will even attack a man who approaches its nest. The female incubates the eggs for 4 weeks while the male feeds the family, and she also begins to hunt when the offspring are 10 days old.

Rufous Hummingbird

SAW-WHET OWL
Aegolius acadicus

The Saw-whet Owl, another owl of the North American continent, inhabits the northern half of the United States and southern Canada, and also occurs in western regions of the United States. It measures about 20 cm and has a wingspan of 45 cm. It is mainly resident in its native locality, but owls from more northerly areas sometimes migrate to the south in winter. Although this owl is relatively common, it is seldom observed because of its nocturnal way of life. During the day, it shelters among the dense branches of tall coniferous trees, setting out after nightfall to hunt for small birds and rodents, and in summer, for insects.

In spring, in the courtship season, these birds make short, repeated whistling cries, and in mid-April the pairs look for hollows abandoned by woodpeckers. They usually build their nests at least 10 m above the ground. The female lays 4 — 6 white eggs and sits on them for 26 — 30 days. The young leave the nest when they are 4 weeks old, but the parents continue to feed them for a further 2 weeks, bringing food to the branches where the young birds perch.

RUFOUS HUMMINGBIRD
Selasphorus rufus

The Rufous Hummingbird is the northernmost representative of the numerous hummingbird species. It covers an extensive zone ranging from north-western California to Alaska. It is most common in Canada, on the fringes of forests, but it is also found in parks and gardens where there is an abundance of flowering plants. This hummingbird reaches a length of 9 cm, and the female differs from the male in having a pinkish-white throat with dark specks. In late summer, the Rufous Hummingbird leaves its home territory to spend the winter in southern Mexico. It returns in spring and nesting begins in May or June. The female builds a strong-walled cup-shaped nest in dense undergrowth or in thick treetops at a height of up to 25 m above the ground. It is made of plant material interwoven with cobweb fibres, and the nesting cup is lined with fine down. The female lays 2 eggs, incubates them for 17 days and rears the brood for 22 days on the nest with no help from the male. This species feeds on nectar, and on tiny flies and spiders, which it pecks off leaves while in flight.

Yellow-shafted Flicker

YELLOW-SHAFTED FLICKER
Colaptes auratus

The Yellow-shafted Flicker is distributed in lightly wooded regions and open countryside where there are scattered groups of trees. It lives in a zone from Alaska to south-eastern Canada and Newfoundland, and spreading south to Florida. The northern populations migrate to the south-east in autumn. This bird is 32 cm long. In the courtship period the male pursues the female in undulating flight, while crying loudly. Both partners chip out a hollow 30 — 60 cm deep with an entrance 7 cm across, in a tree trunk. They make their hollow in isolated trees or near the edges of forests or parks, and usually use the same nest for several years. The female lays 5 — 10 white eggs, which are incubated by both partners for 17 days. The parents feed the chicks for 3 days on the nest and for 10 days outside. Flickers are particularly fond of ants, their larvae and the pupae. In autumn they also eat various kinds of seeds. Migrating birds often stop in coastal forests, in their search for food.

RED-HEADED WOODPECKER
Melanerpes erythrocephalus

The Red-headed Woodpecker is distributed from southern Canada to the eastern and central regions of the United States. It is predominantly resident, only the northernmost populations migrating in autumn to Utah, New Mexico and Arizona. These migratory flights are not regular. This woodpecker is about 22 cm long, and has a markedly bright red head, from which it gets its name. It frequents lightly wooded areas, the edges of forests, and clearings. It flies to open grassland where it feeds on locusts, grasshoppers and other insects, always returning to its perch on a horizontal branch. It complements its meat diet with nuts, fruit and seeds, raiding plantations for the fruit. It also eats eggs and the nestlings of small birds.

The nest is built in a hole in a tree, especially in dead trees. The female lays 4 — 6 white eggs, and both partners share in incubation for 17 days. The young are fed and reared by both parents. They leave the nest after 3 weeks and fly independently.

Woodpeckers of the genus *Melanerpes* are related to

Red-headed Woodpecker

woodpeckers of the genus *Centurus.* About 9 members of the latter genus live in North and South America, the best-known species being the Red-bellied Woodpecker *(Centurus carolinus),* which is found in southern Canada and in the United States.

Red-bellied Woodpecker

YELLOW-BELLIED SAPSUCKER
Sphyrapicus varius

The Yellow-bellied Sapsucker is widespread over almost all of Canada and across the Centre of the United States to Mexico. In autumn, the Canadian populations migrate southwards to spend the winter in Panama, on the Bahamas, and on the islands of Jamaica and Cuba. The sapsucker is about 20 cm long. It frequents the edges of forests, parks, orchards and gardens, where it climbs the trunks of deciduous trees and drills parallel rows of little holes in the bark. It drinks the sap which flows from the holes, and also feeds on insects attracted by the sap. These birds visit orange plantations in the south, bore holes in the fruit with their beaks and suck out the juice.

Nesting season takes place in May and June. The pair makes a hollow 50 cm deep and 15 cm wide, 5 — 20 m above the ground in a dead tree. The bottom of the hole is lined with splinters of wood on which the female lays 4 — 7 white eggs. Both partners incubate the eggs for 14 days, looking subsequently after the young and feeding them on insects and larvae. At the age of 25 days, young sapsuckers leave the nest.

Yellow-bellied Sapsucker

PILEATED WOODPECKER
Dryocopus pileatus

The Pileated Woodpecker lives in eastern regions of the United States, ranging to south-eastern Canada and across southern and south-western Canada to the north-west of the United States. It also lives in Florida. It is a substantial bird, reaching a length of 45 cm. It dwells mainly in large forests, but visits parks and orchards. It feeds on insects, larvae, fruit and seeds.

In spring, each pair chips out a cavity, up to 1 m deep, in a tree trunk, the female doing most of the work. The nest is ready in two weeks, and the female lays 3 — 5 white eggs. Both partners take part in incubation, the male sitting on the clutch at night. The young hatch after 14 days. Their parents feed them on regurgitated insects, especially ants and the larvae of wood beetles. Young woodpeckers leave the nest when they are one month old. They become independent after a further 14 days and scatter over the countryside, often flying many kilometres from their home, to occupy new territories.

HAIRY WOODPECKER
Dendrocopos villosus

The Hairy Woodpecker is distributed throughout almost the whole of Canada, the United States and Mexico. It is about 20 cm long. It is found mainly in mature deciduous or mixed forests, but it frequents large parks as well. The female lays 2 — 6 eggs in the bottom of a hole in a tree, and both partners carry out incubation for 10 — 12 days, the male sitting mainly at night. The hatched young are fed on wood beetle larvae and insects regurgitated by the adults. They begin to fly at the age of 22 days. In autumn, woodpeckers feed on seeds and various fruits.

EASTERN KINGBIRD
Tyrannus tyrannus

The Eastern Kingbird inhabits the North American continent, in a zone ranging from southern Canada to Texas, New Mexico and Florida. In autumn, it migrates to the south and spends the winter in Central and South America, in an area from the coast of Costa Rica to western Brazil, Peru and north-western Argentina. It returns to its North American nesting grounds in spring.
The Eastern Kingbird is about 22 cm long. It is a denizen of thin woodland, parks and gardens, and is also found in bush-covered prairies. In the nesting season, kingbirds fiercely defend their territories. They are very good swimmers and do not hesitate to attack large birds, even raptors, which may approach their nest. In the courtship season, the male often sings, usually after nightfall. In late May, each pair builds a nest of twigs and grass on a horizontal branch. The structure, which is lined with fine twigs,

Pileated Woodpecker

Hairy Woodpecker

Eastern Kingbird

Purple Martin

Tree Swallow

is situated most often at 6 — 8 m above the ground in a tree, though it may be found in a low bush. The female lays 3 — 5 white, brown or grey-spotted eggs, and sits on them for 13 days. The young leave the nest at the age of 14 days, and are fed by both parents. The Eastern Kingbird feeds on insects, catching them on the wing, on leaves, on the ground, or on the surface of the water. In autumn, it also eats plant food and seeds.

Blue Jay

PURPLE MARTIN
Progne subis

The Purple Martin is found in agricultural regions from southern Canada across the west of the United States to central Mexico. It also nests along the Florida coast. It is up to 18 cm long. The female lacks the metallic blue coloration of the male. She is dark brown above and paler below, becoming almost whitish towards the tail. In autumn, the Purple Martin migrates southwards and spends the winter in South America, from Venezuela to eastern Brazil. Early in spring, it returns to the north, ready for nesting time in April. It settles in nesting boxes near human habitations, or builds its own nests in tree hollows or rocky crevices. The nest is made of grass and feathers, and damp soil placed near the entrance assures the right degree of humidity inside. The female lays 3 — 5 white eggs and incubates them for 13 days. The young, leaving the nest after 24 — 28 days, are fed on insects by both their parents. The adults hunt in flight, but also often collect small molluscs from branches.

TREE SWALLOW
Iridoprocne bicolor

The Tree Swallow is distributed from Alaska, through North America to Utah and California in the south. In autumn, it flies south in large flocks to the area from South Carolina to the Bahamas and Cuba, or to Mexico and the mountain areas of Guatemala, returning to its nesting quarters early in spring. This swallow is about 15 cm long. It frequents wooded banks of rivers and lakes, where dead trees abound in which this swallow builds its nest, of grass and feathers. The female lays 4 — 7 pure white eggs and sits on them for 13 days. The young are fed on insects, and leave the nest after 25 days.
When the nesting season is over these swallows gather in large coveys and fly through the region, over marshland, lakes and rivers, skilfully catching insects. In September, the tree swallows join other species of swallows and fly south, though sometimes they stay until early November. In autumn and during their migratory flight, they feed on fruits of various bushes, especially myrtles.

BLUE JAY
Cyanocitta cristata

The Blue Jay is one of the best-known and most beautifully coloured of the North American birds. It reaches a length of about 28 cm. It inhabits the whole of the eastern and central part of the United States, reaching Florida in the south and southern Canada in the north. It prefers dense mixed woodland, both in lowlands and mountains. It also visits gardens near woods. It does not leave its domain in winter, but wanders throughout the region, usually in small flocks of 5 — 20.

The Blue Jay is very cautious, recognizing the first signs of approaching danger. Its loud cry warns the other inhabitants of the forest, and the jay itself noisily seeks safety in flight.

Early in spring, in late March, the pairs occupy their nesting territories. They build a nest of twigs, leaves, bark and dry stalks, preferably on conifers. The nesting cup is lined with fine roots, strips of bark and animal fur. The female lays 3 — 6 pale olive eggs with dark green spots. The young hatch after 17 days of incubation, and both parents feed them on the nest for 3 weeks. When the parents approach the nest, they do so very quietly and they inspect the neighbourhood before settling down.

On leaving the nest, the young stay with their parents and collect forest fruits and seeds. In summer, jays feed on insects, molluscs, worms and occasional vertebrates, and sometimes they take the eggs or nestlings of small birds. The Blue Jay has a wide vocal range and can imitate many sounds as well as the voices of other birds.

COMMON CROW
Corvus brachyrhynchos

The Common Crow is up to 50 cm long, and is found in eastern regions of North America. It inhabits forests and groves along rivers or on small islands. It often builds its nest as early as March, but waits until May in the north. It chooses a suitable tree, most frequently a conifer, and builds its nest 25 m above the ground. The female lays 4 — 8 spotted eggs and sits on the clutch for 18 — 21 days. The male brings food to her and feeds the chicks during the first week. After this time the female helps to feed the brood. Young crows leave the nest after 28 — 35 days.

Common Crow

RED-BREASTED NUTHATCH
Sitta canadensis

The Red-breasted Nuthatch is about 11.5 cm long and the female differs from the male in having grey specks on her head. These small songbirds are found from Newfoundland across southern Quebec to Carolina in the east, and in the west from southern Alas-

Red-breasted Nuthatch

Ruby-crowned Kinglet

Summer Tanager

ka to Colorado and California. In autumn the nuthatch flies south to spend the winter in the south of the United States and in northern Mexico. It lives mainly in coniferous forests, and is abundant in some localities.
The nest is made in a tree hollow, lined with pieces of bark, leaves, hair and feathers. The birds often chisel out holes in the soft, decaying parts of tree trunks. The nest is situated 1 — 20 m above the ground. The female lays 5 — 6 whitish, red-brown-speckled eggs, and incubates the clutch for 14 — 15 days. The young are fed by both parents. At night, the male sleeps in another cavity near the nest, but always sees his mate to her hollow. The fledglings leave the nest when they are 22 — 24 days old, and learn to climb trees. This species feeds on insects and spiders found under the bark of trees. In winter, nuthatches feed on seeds, and sometimes visit bird-tables near human dwellings.

RUBY-CROWNED KINGLET
Regulus calendula

The Ruby-crowned Kinglet is only a small bird, about 10 cm long. It is widespread from Alaska across Canada to the south of the Rocky Mountains. One of its subspecies ranges south to Mexico and to the Guadeloupe Islands. It is resident in pine or spruce forests and it is quite abundant in some localities. In autumn it flies south to Guatemala, and it returns in April.
The nesting season begins in May. Each pair builds a large spherical nest of moss, lichens and webs, lined with fine down. The construction is usually situated at the tip of a branch of a conifer, masked with dense needles. The nest has a small entrance at the top through which the small birds easily slip, to hide from predators. They are preyed upon very heavily by both raptors and carnivores. The female lays 4 — 11 cream-white eggs with purple and russet specks, and incubates the clutch for 16 days. The young are fed on aphids, small caterpillars and spiders, and they leave the nest at the age of 18 — 20 days, though their parents feed them for a further 10 days outside the nest. Flocks of kinglets roam throughout the countryside, and in autumn, before migrating, they feed on small seeds and berries.

SUMMER TANAGER
Piranga rubra

The Summer Tanager lives in open oak and spruce woodland, ranging from Ohio across Nebraska to south-eastern California and northern Mexico. It is also found in Florida. In autumn, tanagers migrate south to spend the winter in the area from central Mexico to Guyana and Peru. The Summer Tanager is 20 cm long, and the female differs from the male in coloration, being olive-green above and yellow below. The male is covered with green and red spots during his first year. This species feeds predominantly on insects, catching them in flight. It hunts bees, wasps and beetles, and collects spiders from cobwebs while on the wing. In autumn, it pecks berries and small seeds.
Nesting takes place in May or June. The female builds a nest of stalks, strips of bark and leaves, and lines it with fine grass. The construction is situated 3 — 5 m above the ground, at the end of a horizontal branch of a tree or tall bush. The clutch averages 3 — 4 pale blue or green eggs with brown spots. The female incubates them for 12 — 14 days, and the male helps her to rear the chicks. The young leave

the nest after 16 days and remain with their parents until they all migrate south.

CARDINAL
Richmondena cardinalis

Originally found on the edges of forests, the Cardinal has now moved into parks, orchards and gardens, and it inhabits agricultural areas where there are trees in which it can build its nest. This songbird has gradually extended its area of distribution north so that it is now resident in Dakota and Illinois. In the south it ranges from Florida over central Texas and Arizona to Mexico. It has been introduced to California, Bermuda and Hawaii. It measures 21 cm and the female differs in coloration from the male. She is predominantly green-brown above, pink-brown below, and has a red crest.
In the north, the nesting season lasts from April to August, but in the south, the birds nest throughout the year, often rearing four broods. The nest is built solely by the female. She makes it of sticks, grass stalks and moss, lining the nesting cup with hair and delicate fibres. When no trees are available, she builds her nest on a tall bush. The nest takes 2 — 3 days to make. Although the male does not participate in building, he accompanies his mate and feeds her. He also announces occupation of the territory by a loud call, consisting of a combination of low- and high-pitched whistling sounds. The female often also sings. She lays 3 — 5 white to greenish eggs with fine red or pale violet specks, and incubates the clutch for 12 — 14 days. The young grow quickly and often leave the nest after 9 — 10 days. The adult birds feed them until they are 4 weeks old.
The Cardinal lives on a variety of seeds, berries and green shoots, and on insects and larvae, finding its food on branches and on the ground. After the nesting season, cardinals merge into small coveys and roam throughout the countryside.

Cardinal
Painted Bunting
Pine Grosbeak

PAINTED BUNTING
Passerina ciris

The Painted Bunting lives in the extreme south-east of the United States, and winters in Mexico, Panama and Cuba. It is about 13 cm long. The female is green above, paler below, and has olive-brown flanks. This bird lives mainly on the edges of forests and on wooded river banks. In the south, it also frequents parks and gardens. The nest is built by the female on a horizontal branch of a bush or low tree.

Solitary Vireo

The female incubates the clutch of 3 — 5 spotted eggs for 12 days. The young are fed at first on insects, and later, on seeds.

PINE GROSBEAK
Pinicola enucleator

The Pine Grosbeak reaches a length of 20 cm, and is a typical denizen of cold northerly regions. Its distribution ranges from Alaska and northern Canada to the west of the United States. It mainly frequents coniferous forests. The nest is built by the female from thin twigs of conifers, stems, moss and lichens, and is softly lined with fine roots. The female lays 4 — 5 bluish, russet-dotted eggs, which she incubates for 13 — 14 days, the male bringing her food. The chicks are fed for two weeks on the nest by both parents, and for 10 days afterwards. This species feeds on the seeds of coniferous trees.

SOLITARY VIREO
Vireo solitarius

The Solitary Vireo inhabits central Canada and most of the United States except for the centre and the extreme south. It also lives in Mexico. It is migratory, flying in autumn to the south of the United States and Mexico, where the local populations are resident. Early in spring, the vireo returns to the northerly nesting grounds in mixed and coniferous woodlands. It seeks sparse woodland both in the lowlands and in the mountains.

The nest is suspended from forked branches of deciduous trees. The vireo chooses thin branches, and its nest is a work of art, woven from stalks and plant fibres, fine strips of bark, spiders' webs and caterpillar silk. The nesting cup is lined with hair and fine grass. The nest is well concealed so it is not easily found by predators. The female lays 3 — 5 whitish, buff-spotted eggs and incubates the clutch for 13 — 14 days. The young are fed on small caterpillars and spiders by both parents. They stay on the nest for 14 days, and the adults look after them for a further 10 days. The diet of adult birds consists of insects found on leaves and twigs. Since the Solitary Vireo feeds largely on the caterpillars of butterflies which are looked upon as pests, it is considered to be a useful species, and therefore enjoys protection.

PARULA WARBLER
Parula americana

The Parula Warbler is a small bird, reaching a length of only 11 cm. Females differ from adult males in coloration, for they lack the buffish pattern. Young birds of both sexes lack the blue shade. This warbler lives in the east of the United States and in the adjacent part of Canada. It also occurs in north-eastern Mexico. It frequents coniferous forests, especially along rivers, both in lowlands and mountains. The nests are made of lichen and tree moss, and hang in massive clumps from branches. These large spherical structures with side entrances look like heaps of moss on the branches, and so escape the attention of predators. The female lays 4, or occasionally 7, eggs having brown and buff dots on a cream-coloured ground. The female incubates the clutch for 12 days. The young are fed on insects, larvae and spiders. The chicks stay on the nest for 12 — 14 days, and on leaving it, they are fed for a further 10 days.

Parula Warbler

MAGNOLIA WARBLER
Dendroica magnolia

This small bird is about 13 cm long, and is distributed from central Canada south-eastwards to the north-east of the United States. In winter it lives in the east of the United States and in north-eastern Mexico. The female differs from the male in having predominantly grey-green upper parts, yellowish underparts and specks along the abdomen. Warblers mainly frequent coniferous forests and they are very abundant in some localities. They build large nests of twigs, needles, grass and webs, lining them with fine roots. The nests are situated in spruces or firs, usually 2 — 5 m above the ground. The female lays 4 whitish eggs with tiny buffish spots. She alone incubates the clutch for 11 — 14 days. The young are fed on insects, larvae and spiders for 14 days on the nest, and for a further 10 days outside it.
Altogether 22 species of the genus *Dendroica* live in North America. They all show similar behaviour patterns.

Magnolia Warbler

BALTIMORE ORIOLE
Icterus galbula

The Baltimore Oriole inhabits the eastern half of North America, from southern Canada to north-eastern Mexico. In winter, it flies south to Colombia and Venezuela. The female differs in colour from the male. She is grey-brown above, yellow-white below, and her tail feathers are olive-yellow.
The Baltimore Oriole is 19 cm long. It lives in thin woodland, often flying to orchards, gardens and plantations. In the nesting season, the flute-like song of the male is heard from the treetops, but the birds are rarely seen, for they are very shy and wary at this time. The nest is built in a forked branch. It is a hanging pouch-like construction, very deep, so that the eggs and nestlings cannot fall out even in strong winds. It is made of long plant fibres intertwined with animal wool, and it opens at the top. The female lays 3 — 6 whitish to pale green eggs, pencilled with greyish lines and specks on the broader end. She

Baltimore Oriole

alone incubates the clutch for 12 — 15 days. The nestlings are fed by both parents who bring larvae and spiders to them for 18 days. When the young begin to fend for themselves after a further 14 days, the adults usually start a second brood. After nesting, families gather in flocks and roam through the countryside, raiding orchards and gardens, and taking ripe fruit, such as mulberries, figs and oranges. They are adept at catching insects on the wing, even snatching wasps. In autumn, flocks of baltimore orioles leave for the south.

EASTERN CORAL SNAKE
Micrurus fulvius

The Eastern Coral Snake is one of the most poisonous of the reptiles. It is an inhabitant of sparse woodland and open bushland in Carolina, Florida, Texas, and in the valley of the Mississippi. It is also found in northern Mexico. It is about 100 cm long. During the day coral snakes hide in underground holes, underneath stones or beneath bark. After nightfall, they leave their shelters and set out to hunt. They glide through crevices under bark and stones, catching small lizards and snakes. They seize the prey either by the head or the body, bite it with their venom fangs, and swallow the paralyzed animal head first.
The female lays a clutch of 6 — 10 eggs under roots or in damp holes in the ground. The young hatch after 3 months, and are about 5 cm long. They are coloured in the same way as the adults, but have lighter shades. They are very active immediately after hatching and soon disperse in the neighbourhood. When alarmed, the Coral Snake hides its head underneath its body and rolls from side to side. The coloured disc which it creates in this way often deters its predators.

BOLAS SPIDER
Mastophora bisaccata

The spider *Mastophora bisaccata* occurs in the southeast of the United States. The female measures about 14 mm, the male only 8 mm. It is found around the edges of forests, in parks and in bush-covered localities. This spider does not make a web, but uses its spinning organs to produce silken fibres about 5 cm long, each terminated by a sticky ball or 'bolas'. The spider holds this weighted thread in one of its front limbs, and swings it round in a circle at high speed when a butterfly or moth is in the vicinity. If it makes contact with the insect, the wings become firmly attached to the sticky blob, and the victim is hauled up. The spider then spins fibres around it, kills it and devours it on a twig.
The female lays her eggs in a special pouch made of web, and attaches it to the bark of a tree. Small spiders hatch out after 10 days, and disperse after a few several days more.

Eastern Coral Snake

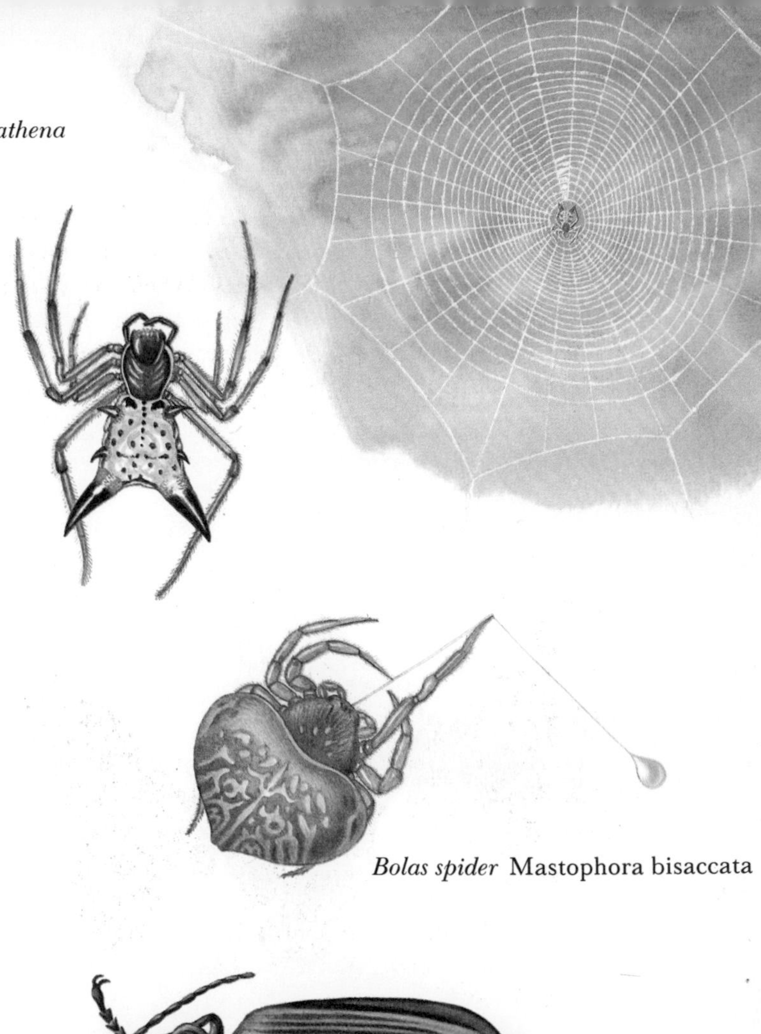

ARROW-SHAPED MICRATHENA
Micrathena sagittata

The Arrow-shaped Micrathena inhabits woods and gardens in the east of the United States. It has a very tough, spiny abdomen. The female is 8 mm long, and the male is slightly smaller. This spider spins large vertical webs among the branches of trees and bushes. It sits in the middle, and as it closely resembles a yellowish flower, insects are attracted to it. It spins fibres around any prey caught in the web, and later eats it.
The female spider lays eggs in a case made of webs and situated in a bark crevice.
Many other members of this genus live in Central and South America.

Bolas spider Mastophora bisaccata

CATERPILLAR HUNTER
Calosoma scrutator

The Caterpillar Hunter is 30 mm long, and has a metallic sheen. This beetle is distributed in woodland areas of North America, where, from May to early July, it frequents mainly deciduous trees. It moves quickly along the branches, pursuing various kinds of caterpillars. It is extremely voracious, and can consume up to 3 large caterpillars every day.
The female lays 100 — 500 eggs which produce equally rapacious larvae within a week. During the fortnight of their development they eat about 40 caterpillars and moult twice. When they are fully grown they leave the trees, and build small chambers in the ground where they pupate. In September or October of the same year, adult beetles emerge, but they remain in the ground until spring, when they move into the treetops. The beetles stay on the trees for 2 months, hunting for caterpillars before spending a further 10 months in the ground. The beetles come out again in spring. The Caterpillar Hunter has a long lifespan of about 5 years.

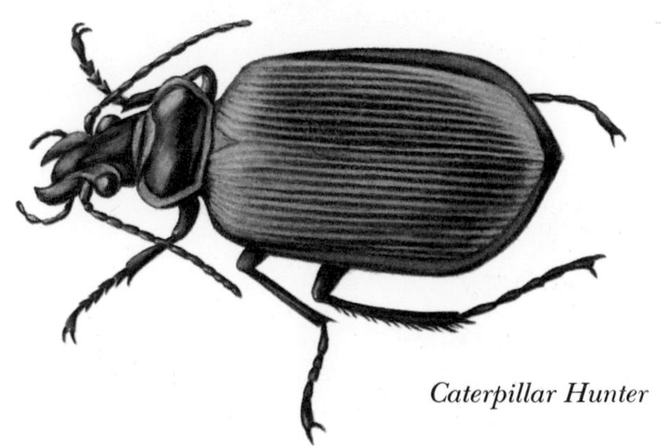

Caterpillar Hunter

Diana Fritillary

Polyphemus Moth

Promethea Moth

American Moon Moth or Luna

White Underwing

DIANA FRITILLARY
Speyeria diana

The Diana Fritillary is one of the most beautiful of butterflies. It has a wingspan of up to 9.5 cm and is characterized by sexual dimorphism — the male being predominantly reddish, while the female is blue-green. This butterfly spends the summer months in clearings in thinly forested areas or on the fringes of humid forests. It flies in a gliding manner, occasionally settling on dry brushwood. This butterfly is distributed from the state of Massachusetts westwards to Illinois.

POLYPHEMUS MOTH
Antheraea polyphemus

The Polyphemus Moth is a beautiful nocturnal moth, having a wingspan of 8.5 — 14 cm. It is distributed in central and eastern parts of the United States and in the adjacent region of southern and south-eastern Canada, while in the south, it reaches north-eastern Mexico. This moth produces two generations annually in the south, but in the north there is only one. The female lays her eggs on various deciduous trees, particularly oaks, elms, white walnuts and maples. The caterpillars have almost globular bodies and feed on leaves. They are green and bear tufts with reddish bases on their body segments. A fully grown caterpillar measures about 7 cm. At this stage, it crawls down to the base of its tree and forms a case out of dry foliage, where it spins a cocoon and pupates. In the south, the second generation overwinter as pupae and the moths emerge in spring.

PROMETHEA MOTH
Callosamia promethea

The Promethea Moth has a wingspan of 6.5 — 10 cm. It lives in the east of the United States, in the north reaching south-eastern Canada, and in the south occurring in Florida and eastern Texas. The male differs from the female in that he is smaller and has a darker coloration. The female is exclusively nocturnal, while the male has the unusual habit of flying out in late afternoon. The female lays her eggs

on magnolias, sassafras and wild sour cherries. The hatched caterpillars are light green, with tiny red, white-rimmed dots on their sides. They bear short orange tufts on the front and hind parts of the body. Before pupating, the caterpillars form shelters of dry, rolled-up leaves, and spin cocoons inside them. The fibres which these caterpillars spin are very firm, and experiments have been carried out to find if they have any industrial applications. The cocoons are usually found at the foot of trees, on bark, or sometimes in the ground.
There are two generations annually.

AMERICAN MOON MOTH or LUNA
Actias luna

The American Moon Moth, or Luna, is a beautiful moth, with a wingspan of 8.5 — 11.5 cm. It inhabits the whole eastern half of the United States and the adjacent part of Canada. It produces two generations. The spring population has pink to reddish wing spots, while the summer form has rich yellow spots. This moth lives in sparse woodland or on the edges of deciduous woods. The female lays her eggs on walnut, maple and other deciduous trees. The caterpillar is light green, with a yellow side line covered with red dots. The back is also red-spotted with white tufts. The adult caterpillar pupates in a case of dry leaves, on the ground, at the base of a tree.

WHITE UNDERWING
Catocala relicta

The White Underwing lives on the edges of forests, and along river banks covered with poplars and willows. It has a wingspan of up to 7.5 cm. It occurs in southern parts of Canada and in north-eastern and central regions of the United States, reaching down to New Mexico. The female moth deposits her eggs on the bark of poplars and willows, the leaves providing food for the caterpillars. During the day, the caterpillars shelter in the bark, coming out only after sunset. They sometimes leave the trees and spend the day resting in fallen leaves, where they later pupate.
Other species of these nocturnal moths, with similar habits, also live in North America.

THE PRAIRIES, MOUNTAINS AND DESERTS OF NORTH AMERICA

Although some of you have never been there, you will probably recognize the countryside we are about to enter, for you see it often in films, especially on your television screens. Most 'Westerns' are set against this background, so if you can follow the action and look at the scenery at the same time, you can get some idea of what the prairies and mountains of North America looked like many years ago. Today, it takes a long time to find an unspoilt site which still looks as it did when the first settlers arrived. For human onslaught rapidly brought about the devastation of the forests, and changed the face of the North American prairies. The vast, grass-covered plains over which the heavy pioneer wagons rolled are no more. Once prairies stretched for 4 000 km from the Canadian province of Alberta to Mexico, and from Indiana west to the Rocky Mountains in a belt 1 000 km wide. It was a sea of grass, interrupted only by islands of low bushes. Now the most prairies have been transformed into huge fields of golden maize, wheat and various other crops. Only a few areas have retained their original character and their typical fauna and flora, and most of these lie within reserves of national parks.

In early May, the prairie becomes bright with anemones, buttercups and erythroniums, later to be followed by composite flowers, such as asters and goldenrods. Buffalo Grass *(Buchloë dactyloides)* and the finer species *Bouteloua gracilis* form a thick green carpet, dominated in some places by Feather Bunch-grass *(Stipa spartea)*. Also commonly found are Northern Dropseed *(Sporobolus heterolepis)* and Western Wheat-grass *(Agropyron smithii)*. Pennsylvania Sedge *(Carex pensilvanica)* and the beard grass *Andropogon gerardi* often shoot up to a height of 1.5 — 2 m. Nodding Wild Rye *(Elymus canadensis)* has a characteristic bluish-green coloration and long ears, and Switch Grass *(Panicum virgatum)* is equally striking. The tallest stems are those of the cord-grass *Spartina pectinata,* reaching a height of 3 m. June is the month of yarrow *(Achillea millefolium),* Daisy Fleabane *(Erigeron ramosus)* with clusters of pale violet flowers, shrubs of the genus *Amorpha,* and the American Liquorice *(Glycyrrhiza lepidota)*. In fact there is a surprisingly large number of colourful flowering plants, enlivening the endless green plains. Twelve species flower in early spring, 40 in spring,

70 in early summer and 40 in late summer. Bushes, such as the Hazel-nut *(Corylus americana)* and the Coral berry *(Symphoricarpus orbiculatus)* also thrive in prairies.

As the white man advanced the animal population had to withdraw from most of the prairie lands. Withdraw is scarcely the right word, for in fact the animals were exposed to wholesale slaughter. Worst affected by the new settlements was the largest American animal, the Bison. In the 17th century, North America from the Yucon to northern Mexico was inhabited by 60 million bison. In autumn, herds several million strong travelled south to seek better pasture, and in spring, they returned to the north. The North American Indians were totally dependent on these bison, from which they obtained not only meat but also hide for making tents, clothes and moccasins, bones for tent supports, tendons for tying poles, and dried dung for heating. The Indians, armed only with bows and arrows and spears, had little effect on the gigantic herds. Later they acquired firearms and horses, and could chase and kill the bison more easily, but the numbers did not diminish because the Indians only killed to satisfy their basic needs. The situation changed with the arrival of the first colonists. They began systematically to kill the bison. They had two reasons for this. The first was to harm the Indians, who were so dependent on the bison, and the second was to improve conditions for cattle breeding. The enormous herds of bison ate huge amounts of grass which were needed for the cattle. Bison meat later provided food for the men working on the Union Pacific Railroad.

W. F. Cody, the famous hunter nicknamed Buffalo Bill, reputedly shot 4 280 bison in 18 months. The period following the year 1869, when the Civil War was over and the railroad finished, turned out to be disastrous for the bison. Thousands of hunters descended on the prairies, killing the animals for pleasure, and sometimes arriving by the trainload. About 10 million bison were exterminated in the next 4 years in the south of the United States, and then the hunters headed north. In 1885, Canada declared the bison a protected species. At that time there were 550 bison left in Canada and 89 in the United States. The almost wholesale destruction was further aggravated by various diseases brought into the North American continent by domestic cattle. At the last possible moment, in 1905, reserves were set up in the United States and the animals were saved. Now 30 000 — 50 000 of them enjoy protection in North American reserves. These huge animals also breed regularly in zoological gardens and there is every hope that the species will be saved for future generations.

Another North American hoofed mammal, the Pronghorn Antelope, experienced a similar fate. At one time its population was estimated at 50 — 100 million, but the antelopes were mercilessly pursued by settlers, and by 1908 the huge herds had shrunk to 19 000 specimens. This shy wary animal was also hunted by the Indians. They did not have the advantage of telescopic rifles, so to bring down an antelope required great skill. The Indian hunters noticed that the cautious pronghorn antelopes showed curiosity towards unusual objects. They fastened brightly coloured rags to bushes where hunters were hiding, and the antelopes were easily killed by their arrows as they approached. Since enjoying protection, these antelopes have multiplied, and now number 350 000 — 400 000 specimens. In some localities numbers have to be regulated to ensure their survival, for in the restricted territories of reserves and protected areas there is only sufficient food for a limited number of animals.

The Coyote is another typical inhabitant of the North American prairies, later exterminated throughout most of its range of distribution. Other denizens of the prairie are a variety of rodents, and rattlesnakes. These poisonous snakes were named after the characteristic 'rattle' at the end of their tails. There have been many opinions on the purpose of the rattle. Some believed it was used to attract females, but the rattlesnake is deaf. Others thought that the snake was warning other animals of its bite, but this snake falls prey to many animals which are unaffected by its venom. In fact, the slow rattlesnake has little chance against fast-moving carnivores and predatory birds, and as a result of its poor eyesight it cannot distinguish friend from foe. There is therefore only one rational explanation. The rattlesnake produces the whirring sound to deter any animal which approaches. Rattlesnakes used to inhabit prairies frequented by herds of bison, and these gigantic creatures could easily have stepped on a rattlesnake and crushed it. When they heard the rattling sound, they stopped and avoided the reptile.

Another legendary animal of the prairies is another hoofed animal, not indigenous to America — the mustang. In the 16th century, the Spanish took horses from Europe to ride in the New World. Many of them escaped, and herds of wild horses which roamed the prairies were the ancestors of the mustangs which became almost indispensable to the North American Indians. Half a century ago, herds of millions could be seen on the grassy plains, but they were exterminated so that now there are a mere 17 000 in the wild. They wander on grass-covered mountain plateaux in the east of the Rocky Mountains, where they are protected.

The extensive mountainous regions of North America are rich in magnificent rock formations. Many mountain slopes are covered with thin woodland, while higher elevations are characterized by stunted scrub and grassy cover. Rhododendrons thrive in some localities; notably the McKinley National Park in Alaska, where they grow at heights up to 3 000 m.

The highest and most massive mountain ranges are found in the west of North America. The Rocky Mountains are the longest mountain system of this area, extending from northern Alaska south to New Mexico. They are composed of many complex systems containing peaks over 5 000 m high. At 6 193 m, Mount McKinley in Alaska, is the highest peak in North America. Another vast mountain system, the Sierra Nevada, extends from British Columbia to southern California. Its highest peak, Mount Whitney, is 4 418 m high. The Coast Ranges, extending along the Pacific, are the third largest mountain belt, and reach heights in excess of 4 000 m.

The alpine localities, which are often stone-covered, have their own characteristic flora. In the Rocky Mountains, the low Alpine Fir *(Abies lasiocarpa)* can be encountered at heights up to 3 000 m, from western Alaska, south to Arizona. Stunted Limber Pines *(Pinus flexilis)* are also common. Hardy deciduous trees like the birches or the Quaking Aspen *(Populus tremuloides)* form extensive cover. The aspens have crooked trunks, deformed from an early age by fierce gales. In spring, these woods abound in flowers, especially anemones, blue columbines, various kinds of sunflowers and Drummond's Mountain Avens *(Dryas drummondii).* Among the high rocks shelter the Narrow-leaved Gentian *(Gentiana algida),* and the clover *Trifolium dasyphyllum* carpets the ground. Rhododendrons with their reddish flowers thrive in early spring in open, sunny localities, while the rocks are red with stonecrop, and milfoil grows in protected places. The Glacier Buttercup *(Ranunculus glacialis)* is found at 4 275 m. Blueberries are common plants of this region, especially the Canada Blueberry *(Vaccinium canadense),* and so are shrubs, such as the Mountain Fetter-bush *(Andromeda floribunda),* and Labrador Tea *(Ledum groenlandicum).*

The mountain regions also have characteristic fauna. They are home to the Rocky Mountain Goat, now protected in national parks. The Bighorn Sheep is another protected species. In 1800, 2 million of these sheep lived in the mountains of North America, but only 7 700 specimens were counted in 1960. This wild sheep was killed both by hunters and by diseases spread by domestic sheep. Flocks of bighorn sheep are now caught, checked and vaccinated before release, so there is every hope that this typical mountain species will be saved.

The California Condor is the symbol of the mountains of California. This bird of prey was rare even before the arrival of Europeans. It was threatened by extinction thousands of years ago, for these birds fed on the carcasses of mammoths and other gigantic animals which became extinct. There are now only about 60 California condors in existence, and they are strictly protected. During the breeding season, environmentalists keep guard near the nesting grounds, to prevent tourists from disturbing these beautiful kings of the mountains.

Among the mountains are the canyons of North America. The largest of these unique natural formations is the Grand Canyon in Arizona. This huge rocky valley is 350 km long, 1.6 km deep and 6 — 12 km wide, and the Colorado River runs through the bottom of it. Bryce Canyon in the state of Utah is the second largest canyon. These valleys are covered with yuccas, agaves, and other plants which enjoy a warm climate.

Desert lands are yet another facet of the American countryside. Vast areas of desert and semidesert ranging from western Texas to the mountains of California, and from Oregon south to Mexico. They include sand and stone-strewn deserts, bush, large cactus-covered tracts, deep gorges and high rocky mountains. All represent various faces of the desert, and all suffer from heat and drought. Vegetation is sparse and discontinuous, and consists mainly of cacti, ranging from miniature species to gigantic ones. Some large species of the genus *Carnegia* reach a height of over 10 m and are thought to be 150 — 200 years old. Cacti, however, have been diminishing in numbers. Cattle breeding in the area has resulted in the loss of the natural vegetation including cacti, and this has given rise to erosion. Furthermore, natural regeneration of the plant cover has been inhibited by the near extermination of coyotes. These animals preyed on rodents, so as their numbers decreased there was a corresponding increase in the populations of fast-breeding rodents which destroyed young plants and seeds. In addition, collectors visited selected places with trucks and vans, and created a cactus rush which upset the natural balance by destroying the cactus cover. Many such localities have now become nature reserves in which the plants are conserved.

In recent years man has striven to pay back his debt to nature and make amends for the errors of the past. In the process he has learned more about wildlife and its environment, and his efforts have resulted in some of the findings and illustrations on the following pages.

STAR-NOSED MOLE
Condylura cristata

The Star-nosed Mole is found in the north-east of the United States and in south-eastern Canada, living in meadows, parks and gardens. This insectivore is about 20 cm long, including its thick tail. Its muzzle is ringed with 22 pink tentacles arranged in a star-like pattern. The outgrowths are covered with cells which are especially sensitive to touch and which enable the mole to detect prey beneath the ground.

The Star-nosed Mole spends most of its life underground, and its eyes are poorly developed. However, eyesight is not important for its way of life, for the mole comes out only rarely and for brief intervals, quickly returning to the network of underground passages surrounding its nest. The mole feeds on worms, insects and their larvae. The female produces a litter of about 5 young after a 3-week gestation period. The young are born blind and open their eyes after 3 weeks. They become independent when they are about 2 months old.

EASTERN AMERICAN MOLE
Scalopus aquaticus

The Eastern American Mole is found throughout the eastern half of the United States. This insectivore is 18 cm long, and is common in gardens and parks, meadows and open woodland. It uses its strong hands to dig tunnels and nesting chambers beneath the ground. It feeds on earthworms and insect larvae. After a gestation period of 3 weeks, the female bears 4 — 5 blind young. These open their eyes at 3 weeks of age, and fend for themselves when they are 2 months old. This mole surprisingly has webbed hind feet, although it is not aquatic. It is a good swimmer but seldom enters water voluntarily.

SHORT-TAILED SHREW
Blarina brevicauda

The Short-tailed Shrew is another insectivore, common in the east of the United States and in southern Canada. It is about 11 cm long and it is one of the most abundant small mammals. It frequents bush-covered and rocky localities, often living near human habitations in cellars and sheds. During the day, it shelters in holes or burrows deserted by other animals, emerging after nightfall to forage for insects and their larvae, earthworms and molluscs. It has slightly poisonous saliva, which paralyzes small invertebrates when they are bitten. A person who receives a bite from a shrew suffers local pain, but is in no serious danger.

In the breeding season, partners locate each other by means of a scented secretion, also used for marking territories. The female bears a litter of 5 — 10 young after a gestation period lasting 3 weeks. The blind and naked young gain their sight after 10 — 13 days and leave the nest when they are 16 days old. They suck their mother's milk for 25 days, and become independent after 5 weeks. They are sexually mature at the age of 3 months.

Star-nosed Mole

Short-tailed Shrew

RINGTAIL or CACOMISTLE
Bassariscus astutus

The Ringtail or Cacomistle inhabits semidesert and desert areas in the south-west of the United States and in northern Mexico. This carnivore is 75 cm long and has a conspicuous, long, striped, bushy tail. It shelters in caves in the daytime, setting out after sunset to comb the neighbourhood for small mammals, birds, lizards, and snakes. It sometimes feeds on the eggs of ground-nesting birds, and about a quarter of its food is made up of fruit. The Ringtail returns to its den before sunrise, and sometimes basks for a short time in the sunshine.
The Ringtail is solitary, forming pairs only for a few days in the breeding season, in February or March. After a gestation period of 2 months, the female gives birth to 2 — 4 blind, sparsely furred offspring, which open their eyes after 31 — 34 days. The young leave the nest when they are 35 — 40 days old. They are suckled for 2 months, but begin to feed on prey brought by their mother from the age of 35 days.

COMMON or STRIPED SKUNK
Mephitis mephitis

The Common or Striped Skunk inhabits almost the whole of North America except for the northernmost regions. It frequents the edges of forests and shrubby meadows. This marten-like carnivore is about 1 m long. During the day it shelters in a rocky crevice, hollow stump or hole beneath the roots of a tree or bush. After nightfall, it comes out to forage in its territory, moving nimbly on the ground and probing every crevice. It preys on small rodents, lizards, snakes, insects and molluscs, and occasionally eats fruit. When threatened by a coyote or a dog, the skunk sprays its enemy with a yellowish secretion from vents under its tail. The strong-smelling liquid sprays over a distance of several metres, and this usually discourages the adversary. The affected animal cannot get rid of the foetid smell for several days, an experience from which it learns to avoid skunks in future.
Following a gestation period of 61 — 64 days, the female bears 3 — 8 young. These are born blind, and gain their sight after 20 — 28 days. Their mother looks after them carefully and suckles them for 6 —

Eastern American Mole

Ringtail or Cacomistle

Common or Striped Skunk

Spotted Skunk

8 weeks. At the age of 35 days the young leave the burrow to play in front of it. They become independent at 10 months and seek their own territories. Despite its efficient means of defence, skunks are caught in large numbers, especially by birds of prey. They are also hunted for their fur, and for this reason are now bred on farms.

SPOTTED SKUNK
Spilogale putorius

The Spotted Skunk is about 55 cm long. It is found throughout the United States except the extreme north and north-east, and it also lives in northern Mexico. In all these regions this carnivore is common. It frequents dry rocky places, bush-covered prairies and sites near farms, where it settles in storehouses, sheds, or other suitable buildings. During the day, the skunk sleeps in its den. It comes out after sunset to search for insects, small rodents, lizards and snakes. It also takes fruit from gardens and orchards. Like other skunks, it has a scent gland in vents under its tail, and it sprays attackers with a foul-smelling liquid.
The female bears 3 — 5 blind young after a gestation period of 2 months. The young open their eyes after 20 — 30 days and leave the den at the age of 35 days. Their mother suckles them for 7 — 8 weeks. Young skunks become independent after 10 — 11 months, but they are fully grown when they are 5 months old.

Coyote

COYOTE
Canis latrans

The Coyote is one of the best-known animals of the American prairies. It is about 100 cm long. It used to range from Alaska south to Costa Rica, but it has been exterminated in many places and has only survived in some numbers in the west of the United States and in coastal regions of Mexico. The Coyote lives in family groups which gather into packs during the winter. It hunts after nightfall, sometimes catching small vertebrates, but mostly feeding on carrion and occasionally eating fallen fruit. The gestation period of this carnivore is 60 — 65 days. The female bears 3 — 4 offspring in May, in a den dug in the ground, or in a cave. The young are born blind and gain their sight after 8 — 12 days. They are suckled for 8 — 10 weeks, but their parents often feed them meat from their fourth week of life.

PUMA or MOUNTAIN LION
Felis concolor

The Puma or Mountain Lion is a large member of the cat family, reaching a body length of 2.5 m and a weight of 100 kg. This carnivore is distributed all over North, Central and South America, forming several subspecies differing in size and coloration. The silver-coloured Canadian Puma is the largest, while the South American populations are smaller and redder. It is known by various names, including silver lion, cougar and panther. The Puma was once abundant, but it has been exterminated in many places. In the North American continent, it is now common only in north-western Canada and in the west of the United States.
The Puma inhabits rocky localities and sparse woodland, especially in the mountains. It usually hunts after nightfall, but it sometimes hunts during the day, lurking on trees or rocky outgrowths, and pouncing on its prey from above. It also climbs trees to catch roosting birds. Pumas prey mostly on va-

Puma or Mountain Lion

rious species of deer, particularly on young animals. They also catch rodents, rabbits and hares.
In the rutting season pumas make long whistling sounds, and the female screeches. After a gestation period of 92 — 96 days, she bears 1 — 5 young. In the north, pumas have their young in summer, but in the south breeding takes place in any season. The spotted young are born blind, opening their eyes after 9 — 10 days. They leave the den at the age of 35 days and begin to gnaw meat brought by their mother when they are 45 — 50 days old. They are suckled for 4 — 6 months, and remain with their mother for 2 years. Females mature after 2 years, males a year later.

BLACK-TAILED PRAIRIE DOG
Cynomys ludovicianus

The Black-tailed Prairie Dog inhabits the vast grasslands of the centre of the United States, from Montana south to Arizona and Texas. It is not a dog, but is a rodent about 35 cm long, which lives in large communities. A single colony may number up to several thousand, and in places where prairie dogs live, the ground is covered with holes. The numerous exits to the underground burrows are protected from rain by cone-shaped mounds of damp earth, which the prairie dogs maintain by stamping on the soil and flattening it with their muzzles. If a rattlesnake enters the burrow in search of their young, the adults bury the snake and stamp the earth down so hard that the intruder cannot get out again, and so dies. Deserted burrows are often taken over by burrowing owls.

Black-tailed Prairie Dog

Greenland Collared Lemming

Prairie dogs feed on grass and the green shoots of plants and shrubs. Several members of each colony are constantly on guard, and warn the others if a predator, such as a bird of prey, approaches. Their warning call is a sharp whistle. Despite their caution, many young prairie dogs do fall prey to raptors and small carnivores.
In spring, the female gives birth to 4 — 6 blind young after a gestation period of 27 — 33 days. The young open their eyes after 33 — 37 days, and immediately go outside the burrow to bask in the sun. Their mother suckles them for 7 — 8 weeks.

GREENLAND COLLARED LEMMING
Dicrostonyx groenlandicus

The Greenland Collared Lemming is one of the most abundant of the rodents of the far north, being found mainly in the tundra of Alaska and northern Canada. The number of lemmings in the wild varies regularly, large increases in population taking place usually once every 4 years. This is conditioned by the quantity of food, rainfall and winter climate. There are a number of factors affecting the population explosions, but the principal causes remain unknown even to experts. Whatever the reasons, the lemmings then leave their homes in millions. They do not travel in order to drown in seas or lakes, as they are reputed to do, but to find new territories and new sources of food. They are not stopped by any obstacle in their path and may drown while swimming across lakes and rivers. Foxes, wolverines, raptors and owls also help to reduce their numbers, and themselves have larger litters in the years when there are huge numbers of lemmings, for then there is no shortage of food.
Lemmings are nocturnal. They build surface corridors or 'runways' on the ground, nimbly moving along them. They do not hibernate in winter, moulting instead into white coats and being active even in the coldest weather. They build spherical nests of grass and leaves beneath snowdrifts, and feed on the bark of trees and bushes, dry grass and leaves. In summer the nests are made underground, near to the surface, or sometimes in low bushes.
After a gestation period lasting 18 — 21 days, the female bears 10 — 12 young, sometimes 3 times a year. The young are sexually mature at the age of 3 weeks, which is one of the reasons why lemmings tend to overbreed.

Eastern Cottontail

EASTERN COTTONTAIL
Sylvilagus floridanus

The Eastern Cottontail lives in the east of Canada and the United States, in central America and in the north of South America. This common rabbit measures about 30 cm and weighs up to 2 kg. It frequents shrubby localities, forest margins and sparse forests. It is nocturnal, leaving its shelter among clumps of tall grass or beneath thickets after nightfall to feed on grass, leaves, shoots and bark.
The female gives birth to 4 — 7 young after a gestation period of 26 — 31 days, though litters as small as 1 or as large as 14 are sometimes produced. The young are born blind, gaining their sight after 8 days. They are weaned at 17 — 18 days and be-

come independent when they are 25 days old, at which time they leave their mother. They mature after 5 — 6 months. Some females bear 5 — 7 litters a year. The Florida Cottontail falls prey to wild dogs and cats, martens and birds of prey.

AMERICAN BISON
Bison bison

The American Bison, also called the American 'Buffalo', is the largest of the American animals. Old bulls attain a body length of 3 m and a height at the shoulder of 190 cm. They weigh 1 000 kg or more. The cows are smaller and less thick-set, and have more delicate horns. This species occurs in two subspecies; the prairie and wood bisons. The latter is more heavily built and lives in Canada.
The Bison lives in large herds made up of many small groups of 20 — 25, each led by a strong bull. Young bulls often have their own communities of 2 — 12. After a 270 — 285-day gestation period, the female gives birth to a single calf, or sometimes twins. The calf stays with its mother until the next winter. Bison have a life expectancy of 25 years. They are hardy and undemanding animals, active both in daytime and at dusk. They graze on tough grass and in winter they feed on dry grass and lichens, dug out of the snow with their horns.
Their enemies in the wild include wolves, packs of which can chase a stray young bison to exhaustion.

Caribou or Reindeer

CARIBOU or REINDEER
Rangifer tarandus

The Caribou or Reindeer is nowadays confined to Alaska and Canada. Although it formerly inhabited the region around the Great Lakes and the states of New York, Washington and Idaho, it no longer occurs in the United States. The Caribou frequents

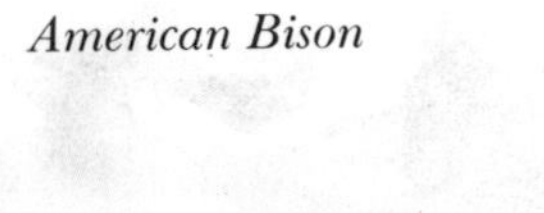

American Bison

Rocky Mountain Goat

open country, especially tundra. The bull reaches a weight of up to 150 kg, the female being smaller and having less massive horns. Caribou spend their life in herds, only old males being sometimes solitary. Rutting takes place in autumn, when the bulls engage in fights and the winners gather herds of 10 — 20 cows. After returning to their summer quarters in the north, the female gives birth to a single calf, or occasionally to twins or triplets. The gestation period lasts about 220 days. The calf is born completely developed, and is able to follow its mother the day after it is born. This deer feeds on grass, sedges and the shoots and leaves of birches and willows. In winter, it eats mosses and lichens, dug from beneath the snow.

ROCKY MOUNTAIN GOAT
Oreamnos americanus

The Rocky Mountain Goat inhabits rocky areas above the timber line in south-western Alaska and western Canada, and south to northern Montana in the United States. The male is 1.5 m long and 1 m high at the shoulder. The female is smaller. This species has very long, thick, pure white hair, which protects it from the bitter weather and from predators, for in winter the goats cannot be seen against the snow-covered slopes. Their hair is dry, and quickly absorbs rain, though rainfall is sparse in the areas where they live. Rocky mountain goats frequent the edges of snowfields at altitudes of 4 000 m, descending to lower levels only in winter. From spring to autumn they feed on grass, leaves and twigs of stunted shrubs. Lichens and moss form the bulk of their winter diet.
The Rocky Mountain Goat lives in herds led by robust males. The rut takes place in November. After a 165 — 185-day gestation period, the female bears a single young or twins in late April or early June. The young mature at 2.5 years of age. The Rocky Mountain Goat is protected in a number of National Parks and reserves.

Bighorn Sheep

Pronghorn Antelope

BIGHORN SHEEP
Ovis canadensis

The Bighorn Sheep lives in rocky mountain ranges in the west of the United States and in south-western Canada. Having been exterminated in many places, it is now most common in the Californian mountains, where it enjoys protection in reserves. The male measures up to 180 cm, stands 1 m high at the shoulder, and weighs over 150 kg. The female is smaller. The rams have massive horns, up to 70 cm long, while the females have shorter, thinner horns. The Bighorn Sheep dwells on steep, virtually inaccessible slopes. The rut takes place form mid-November to the end of December. At that time, the sound of clashing horns echoes through the mountains, as the rivals fight each other. After a gestation period of 180 days, the female bears a single young, or sometimes twins. The lambs are extremely active, nimbly climbing the slopes and following their mother the day after they are born.

PRONGHORN ANTELOPE
Antilocapra americana

The Pronghorn Antelope is native to the American prairies and grasslands. It once covered the whole of the west and central part of the United States, reaching to southern Canada and northern Mexico. However, its numbers were drastically diminished by hunting, and its area of distribution consequently became reduced over the years.

The Pronghorn is not a true antelope, and is the only representative of its family. It is 150 cm long, 90 cm tall at the shoulder, and weighs 55 kg. Both sexes have unusual forked horns, unique among horned mammals. They are 30 cm long, are flattened from sides, and differ from other types of horns in that they have a horny sheath made of a layer of modified hair covering the bony core. The sheath is shed annually after the breeding season.

The Pronghorn lives in small groups, usually composed of one male and 2 — 8 females. Herds consisting only of males are frequent, and old males are often solitary. The herds live on the prairies and on grassy hills, near water. During the summer, they roam in small bands in search of food, gathering in larger groups in winter. They feed on grass, plants and the leaves of shrubs. When threatened by wolves or raptors, they flee rapidly and usually avoid capture. The Pronghorn is the fastest land animal, after the Cheetah and can reach a speed of 80 km/hour. It can raise the longish white hairs around its tail, when it senses danger, and so warn other members of the herd. The signal can be seen over long distances.

In September, mature males fight for groups of females. After a gestation period of 8 months, the female finds a sheltered spot and gives birth to twins. The young follow her the day after they are born.

Sandhill Crane

Nene or Hawaiian Goose

SANDHILL CRANE
Grus canadensis

The Sandhill Crane is found in open grasslands with sparse woods and scrub. It is distributed from northern Alaska across Canada and the United States to Mexico and Cuba. It forms 6 subspecies in this vast territory. Those subspecies occurring in Cuba, Florida and the Mississippi valley are resident, the other three are migratory, spending the winter in large colonies from California and Texas to Mexico. Some 50 000 cranes winter in Texas, and the total population in North America numbers about 200 000 birds. The Sandhill Crane measures up to 95 cm long and has a wingspan of 2 m. Migrating birds fly 120 — 300 m above the ground in V-shaped formations, usually in flocks of 40 — 200. In spring, they return to their native areas and form pairs which look for nesting territories. During the courtship period the pairs dance, hop around each other, flap their wings and make trumpeting sounds. They build nests 50 cm across and 20 cm high of stalks and sticks, on the ground, usually in tall grass. In the north, they nest in May, in southern regions in April. The female lays 2 pale olive, brown-spotted eggs into the grass-lined nesting cup and both partners undertake incubation for one month. The chicks leave the nest a few hours after hatching, and wander with their parents in the neighbourhood. The parents bring them food during the first few days. Sandhill cranes feed on insects, molluscs and small vertebrates, and on seeds, berries and green plants. After nesting, the birds merge into flocks.

NENE or HAWAIIAN GOOSE
Branta sandvicensis

The Nene or Hawaiian Goose is about 60 cm long. At one time it was common in the Hawaiian Islands and in the adjacent island of Maui. However, it became virtually extinct in the wild. Hunters shot these tame geese in their hundreds, and by 1930, all the nenes on Maui had been killed. By 1949 only some 30 specimens remained on Hawaii. At this point stock from Hawaii was sent to the headquarters of the Wildfowl Trust at Slimbridge in England. Here the Nene was successfully bred and several hundred birds were re-introduced to Hawaii. The rare goose was saved. It is a resident species, living on open mountain slopes, at heights of 1 500 — 2 500 m. Nesting takes place from November to March, the female laying 6 — 8 eggs in a sparcely-lined depression. She incubates the clutch for one month, and the goslings are cared for by both parents.

CALIFORNIA CONDOR
Gymnogyps californianus

The California Condor used to range along the west coast of North America from the state of Washington to the north of Baja California on the boundary between the United States and Mexico. It is now restricted to a small territory along the Californian coast, occasionally wandering to southern Oregon and Arizona outside the nesting season. Although this bird of prey was never abundant, it was much sought by hunters, and its eggs were collected. This beautiful raptor reached the verge of extinction and was only saved by strict protection. Nevertheless, only 60 specimens live at present in the wild.
The California Condor measures 125 cm and has a wingspan of over 3 m. It weighs 8 — 14 kg. It occurs high in the mountains, among rocks, from which it takes off to fly above the mountain valleys and river banks. It seeks the carcasses of large game or the remains of prey killed by large carnivores. It sometimes travels as far as 100 m in search of food, its daily intake of meat being about 1 kg. Condors hunt live prey only when food is in short supply. Then they attack small vertebrates, such as ground squirrels, killing them with their claws.
The California Condor perches for hours on rocky platforms or dry branches, sometimes remaining in the same position for more than 15 hours. It usually roosts from 1 p. m. through to 8 a.m. of the following day, foraging for food in the morning. A great deal of time is devoted to preening, and basking in the sun is also a favourite occupation of this condor.
The nesting season lasts from January to May. The birds build a nest in a cave, or sometimes in the hollow of an old tree. The nest is usually situated 500 — 1 500 m above the sea, sometimes only a few kilometres from the sea coast. The female lays one, or occasionally two bluish or green-white eggs, in January or February, and incubation is carried out by both partners for 42 — 50 days. The young condor

California Condor

grows slowly and the parents look after it with great care. At first, they bring it food twice a day, in the morning and in the evening, but later it is fed just once a day. At the age of 3 months, the young, still flightless bird, leaves the nest and moves to neighbouring rocks, where its parents continue to feed it. It is capable of flight after about 6 months, but becomes a good flier only after 10 months. Until that time, it is still regularly fed. Because the offspring take so long to rear, condors nest only once every two years. Young birds have grey-black plumage and mature after 3 years of age.
The California Condor makes only hissing or harsh rattling sounds. Some birds live to be 50 years old. Sadly it is one of the birds threatened by extinction.

AMERICAN KESTREL or AMERICAN SPARROWHAWK
Falco sparverius

The American Kestrel or American Sparrowhawk is a small raptor measuring 25 — 29 cm. It has a large

American Kestrel or American Sparrowhawk

area of distribution, from Alaska across North America, and south to the West Indies and Chile. Fifteen subspecies live in the wild. This kestrel frequents bush-covered localities with solitary trees, deserts, semideserts and parks in large cities. It perches on branches of trees or tall bushes, on ledges, and on telegraph poles from which it takes off to catch its prey. In hot deserts it hunts only early in the morning and before nightfall, sheltering in the shade during the day. It is a fast and skilful flier, and can catch insects on the wing, especially locusts and grasshoppers flying near the ground. In late afternoon it also catches small bats, and occasionally takes small birds, lizards and frogs.

For most of the year kestrels live in large flocks which fly south for the winter. They form pairs in the nesting season, which is in May and June in Canada, mid-May in the United States, March in Florida, and September and October in Chile. They find hollows in trees or cacti, or even in nests deserted by other birds of prey. The female lays 4 — 5 eggs at intervals of 2 — 3 days. The eggs are cream-coloured and densely covered with red and brown spots. The female incubates the clutch for 29 — 30 days, while the male feeds her, and sometimes takes over incubation at night. At first the young are fed by the female, the food being brought by the male. Later he also feeds the young on insects. Kestrels leave the nest when they are 30 days old, and the families stay together during the winter.

Greater Prairie Chicken

GREATER PRAIRIE CHICKEN
Tympanuchus cupido

The Greater Prairie Chicken is about 45 cm long, and inhabits grassy prairies with a thin cover of scrub. It used to be very common, but is no longer found in any number except in southern Canada, northern and central parts of the United States and in some localities in the south of the United States. The males have tufts of feathers on the sides of the neck which stand vertically when the birds display. They also have orange pouches along the neck, which they inflate in the courtship period. At the same time they extend and raise their tail feathers, hang down their wings and make bubbling sounds.

The female makes a shallow depression in a clump of grass or under a low bush, and lays 8 — 12 olive, brown-spotted eggs. She incubates the clutch for 23 — 24 days, and looks after the chicks by herself, taking them to where there is food, and sheltering them under her wings at night. This bird feeds on insects, molluscs, worms, seeds and green plants, and it often digs in the ground in search of prey.

SAGE GROUSE
Centrocercus urophasianus

The Sage Grouse is found in the north-west of the United States and in a small adjacent area of Canada, frequenting bush-covered localities. The male is 70 cm long, but the female is only 55 cm long and lacks the black pattern on the neck and throat. During the courtship season, groups of males gather to take part in nuptial displays. They raise their neck feathers to produce a large white collar in which their two red throat pouches shine like two bright balloons, and their bubbling calls can be heard as far away as 300 m. Success in courtship depends on the position in the hierarchical structure of each male. The winning male takes about 74 per cent of the available females into his harem. The second-ranked male gets about 13 per cent of the hens, and the other subordinate males share the rest of the females accordingly. The males do not participate in incubation or in the rearing of the brood. The female makes a shallow depression under a bush, lines it with dry leaves and grass, and lays 6 — 8 olive or greenish eggs which she incubates for 22 days. She then protects the young, which begin to fly after 2 weeks.
After the young leave the nest, families gather in large flocks of up to 100 birds. The Sage Grouse feeds on invertebrates such as beetles, and on seeds, berries and green plants. It drinks copiously twice a day.

Sage Grouse

BOBWHITE
Colinus virginianus

The Bobwhite is a small fowl-like bird, about 24 cm long. It lives in bushy localities, fields and sparse pine woods in the eastern half of the United States. It also occurs in a small area in the north-west of the United States, and in north-western Mexico. It forms pairs in the nesting season, at which time the male's whistling trisyllabic call is often heard. The female lays 8 — 20 white eggs in a sparsely lined depression under a thick bush or in a clump of grass. The male sometimes helps to build the nest, but the female incubates the clutch by herself for 21 — 23 days. The chicks feed on insects and their larvae and on ant pupae found in the pine woods. Later they collect molluscs and seeds and peck at plants, and they begin to fly at the age of 3 weeks. The male takes part in the rearing of the brood. Some pairs nest twice or three times a year. In winter, bobwhites live in flocks of about thirty.

Bobwhite

Barn Owl

BARN OWL
Tyto alba

The Barn Owl is widespread over almost all of the United States. It is 35 cm long and has a wingspan of 110 cm. It inhabits rocky sites, and its hoarse voice is often heard near human settlements. It nests in a rocky crevice, or behind a beam in a barn, or in some similar place, using no nest material. The female lays 4 — 6 white eggs, and sits on the clutch for 30 — 32 days, while her mate feeds her. The offspring, which leave the nest when they are 52 — 58 days old, are reared by both parents. The Barn Owl mainly hunts rodents, but it also catches bats, amphibians, small reptiles and occasionally small birds.

Snowy Owl

SNOWY OWL
Nyctea scandiaca

The Snowy Owl inhabits arctic regions of Alaska, northern Canada and the adjacent islands. It winters regularly all over Canada and sometimes in the north of the United States. It measures 54 — 66 cm and has a wingspan of about 135 cm. The adult male is almost pure white, but the female is densely spotted. During the nesting season the Snowy Owl lives in open country with low scrub. Each pair occupies a territory covering over 2 square kilometres. The nest is built on the ground, in a depression scantly lined with feathers. The female lays 4 — 6 white eggs, but when, as sometimes happens, there are huge populations of lemmings, the clutch may number up to 10 eggs. Lemmings are the Snowy Owl's main source of food, and at times of shortage, the owl does not breed. Nesting takes place from April to June, and incubation lasts 32 — 34 days. The female starts sitting when the first egg is laid, and the young hatch out gradually. The youngest ones usually perish in competition with their larger siblings. Young owls stay on the nest for 57 — 61 days, until they are capable of flight. The parents bring lemmings for them, as many as 4 a day for each chick. One Snowy Owl can consume 600 — 1 600 lemmings in a year! In winter, when the lemmings are living beneath snowdrifts, the owls migrate to the south where they catch other rodents and small birds, hunting even during the day. During the summer they are obliged to hunt when it is light, for at that time inside the Arctic Circle there are few hours of darkness.

BURROWING OWL
Speotyto cunicularia

The Burrowing Owl is distributed in the prairies and semideserts of south-western Canada, the western half of the United States, and Mexico. It also occurs in South America, but not in the Amazon region. It is resident in southern areas, and migrates from northerly regions to the south in autumn. It is a small owl, only 23 cm in length and with a wingspan of 55 cm. The Burrowing Owl usually lives in colonies made up of 10 — 12 pairs which often nest near each other. The birds look for burrows deserted by prairie dogs or other rodents, or dig out underground passages. The nesting chamber is up to 1 m below the surface, at the end of a corridor 3 m long. The nesting cup is lined with grass. Here the female lays 2 — 12 white eggs, which are incubated by both partners for 4 weeks. The young are also fed by both adults, hunting during the day as well as at night. These owls feed on large beetles and other insects, small rodents and occasional birds or frogs. During the day, they bask in the sun in front of their burrows.

Burrowing Owl

ROADRUNNER
Geococcyx californianus

The Roadrunner inhabits deserts and semideserts ranging from the south-west of the United States to central Mexico. This interesting bird measures about 60 cm, of which half is tail. It has long legs and it runs very quickly over short distances reaching a speed of 35 km/hour. It also takes long jumps, and can fly quite well. The Roadrunner moves briskly through the undergrowth or across extensive sandy areas thinly covered with grass. It searches for insects, molluscs, small birds and rodents, and pursues small lizards and snakes, including poisonous rattlesnakes. It seizes a snake or iguana behind the neck with its long, sharp beak, and smashes it against a stone before swallowing it head first. It also breaks hard snail shells on stones, and then swallows the soft mollusc bodies.
The nesting season is in April or May. The birds build a nest of twigs on a low bush or stunted tree, and line it with sloughed snake skins. The female lays 4 — 10 white eggs and sits on them by herself for 19 days. The hatched young are fed on insects and later on small vertebrates. When the young leave the nest the adults usually rear another brood. The Roadrunner makes cooing sounds like a pigeon, but sometimes clacks its beak like a stork.

CHIMNEY SWIFT
Chaetura pelagica

The Chimney Swift is a common bird of North American cities. It is 12 cm long, and originally nested in tree hollows, but it adapted rapidly to the advance of civilization and moved to the towns and cities, where

Roadrunner

Chimney Swift

it nested in a very unusual place — inside tall school or factory chimneys! The swift first covers the inner, smooth wall of the chimney with its sticky, fast-hardening saliva. It breaks off twigs from trees as it flies and takes them to the chimney, glueing them together with saliva and forming a small bowl-shaped nest. The female lays 3 — 6 eggs which are incubated by both parents for 19 days. The young are fed on insects caught on the wing. The nest soon becomes too small, and the young hang themselves outside it or on the wall. They are capable of flight after 35 days, and set out for their first short trips. The families then gather in large flocks and travel together to their winter quarters in South American cities where they again roost in chimneys.

Ruby-throated Hummingbird

RUBY-THROATED HUMMINGBIRD
Archilochus colubris

Hummingbirds are among the best-known of American birds. One of them, the Ruby-throated Hummingbird, is distributed in the eastern half of North America from the Great Lakes to Florida. It frequents open countryside covered with trees, especially gardens, parks and orchards. Like other hummingbirds, it enjoys bathing in leaves containing water. It is about 9 cm long and, unlike the male, the female has a white, black-spotted neck.

Hummingbirds are solitary for most of the year. In the nesting season the male seeks a partner, but leaves her after mating to look for another. The courting male flies around in wide arcs, pursues his partner, and makes buzzing sounds with his wings. He often emits a high-pitched, whistling call. The female builds a cup-shaped nest of moss and spiders' webs, intertwined with lichens. It is situated in a tree or bush, in a forked branch, usually 3 — 5 m above the ground. In May or June, she lays 2 white eggs and sits on them for 16 days. She feeds and rears the offspring alone, bringing them nectar and tiny insects. Young hummingbirds leave the nest when they are 22 — 24 days old and learn to find their food. They feed on nectar from flowers and catch insects in flight. In autumn, they leave their homes and migrate as far as 800 km to the south, spending the winter in Mexico, Central America and West Indies.

HOUSE WREN
Troglodytes aedon

The House Wren is common on both American continents. It ranges in a belt from southern Canada across the United States to Tierra del Fuego at the very tip of South America. It is found in about 30 subspecies throughout this enormous territory. The northerly populations migrate southwards in autumn, and the birds from the southernmost regions return to the north. The North American populations spend the winter in California, Texas and Mexico. The House Wren attains a length of about 12 cm, and lives in gardens and parks. In spring, the male takes up occupation of a territory and fiercely defends it. The nest is usually made in a nesting box, under roofs or in hollows and stumps. A discarded can is often used. The male makes the nest of dry twigs, grass, feathers and spiders' webs, while the female lines the nesting cup with little pieces of bark or down. She incubates the clutch of 6 — 8 white, pink or red-spotted eggs for 14 — 15 days. She usually feeds her offspring alone, but she is sometimes assisted by the male. Wrens feed on insects, their larvae and spiders. The young leave the nest when they are about 16 days old.

CACTUS WREN
Campylorhynchus brunneicapillus

The Cactus Wren inhabits arid regions covered with cacti and thorny scrub. It is a large wren, about 22 cm long and occurs predominantly in desert areas of southern California, Nevada, Utah, New Mexico, Texas and northern Mexico. Insects, especially beetles, form the bulk of this wren's diet, but it sometimes pecks berries and green plants and sometimes eats seeds. The nest is made among the sharp spines of the cholla cactus or in thorny shrubs. It is a spherical structure woven chiefly by the male, from twigs, thorns and grass. It is lined with fine feathers by the female. The male builds several other nests, sleeping in them himself. The clutch averages 4 — 7 white eggs with dark brown spots. Incubation lasts 15 days, and the rearing and feeding of the young is undertaken mostly by the female. The male occasionally helps by bringing beetles or locusts. The young leave the nest at the age of 17 days.

CATBIRD
Dumetella carolinensis

The Catbird is distributed from southern Canada to Texas. In winter, it migrates to Central America and to the Antilles, returning to its nesting grounds in April. It is 23 cm long and inhabits the banks of

House Wren

Catbird

Cactus Wren

American Robin

rivers and lakes in localities overgrown with bushes. The courting male follows his mate, perches on an elevated branch and raises his tail feathers, decorated below by a russet spot. In May, or in June in northern habitats, both partners build their nest. It is made of roots, grass stalks, wool and paper, and is lined with fine plant fibres. The female lays 4 — 6 blue-green eggs and incubates them alone for 12 — 13 days. The young are fed on insects, worms and small molluscs by both parents. The insects are often caught over the water. Young catbirds leave the nest at the age of 2 weeks, but they are fed for a further 10 days.

Eastern Bluebird

AMERICAN ROBIN
Turdus migratorius

The American Robin is one of the best-known North American birds. It reaches a length of 25 cm, and the female has paler and less bright colours than the male. This bird frequents open locations with bushes and scattered trees, on the edges of forests in parks and in gardens. It is widespread from north-western Alaska across Canada and south to South Carolina, and to Guatemala in South America. In autumn, robins flock together and migrate south, the northernmost populations wintering in Nebraska and Wyoming. Large flocks settle for the night in marshes and extensive reed beds. During the day, smaller coveys roam throughout the countryside in search of berries, worms and insects. They return to their native land in April and form pairs. The males fight for occupation of the best territories. In May, they build a bowl-shaped nest from grass stems and mud, lining it with soft plant fibres. They rear a second brood from July to August, this time in deciduous trees. The female lays 3 — 5 blue-green eggs and sits on them for 13 days. The young are fed by both parents, on insects, larvae and worms. At 2 weeks of age, young robins leave the nest, still unable to fly. They hide in clumps of grass and among stones, where the parents bring them food, and during the following week, the young robins learn to fly and find food for themselves.

EASTERN BLUEBIRD
Sialia sialis

The Eastern Bluebird belongs to the thrush family. It is about 18 cm long, the female being brownish above, paler below, and having a white throat. This species is distributed from eastern Canada southwards to Florida and westwards as far as Montana. From Florida, it has spread across western Mexico to Honduras. The Canadian populations winter in the United States south of the Great Lakes. Bluebirds

inhabit open scrubland, parks and gardens. In the nesting season, in May and June, the male seeks a vantage point on a branch up to 35 m above the ground, and sings loudly. The nest is made in hollows deserted by woodpeckers or in nest boxes. It is composed of grass stems and is situated from 50 cm to 10 m above the ground. A clutch of 4 — 6 pale blue eggs is incubated by the female for 13 days. The young are fed on insects and their larvae. Young bluebirds leave the nest when they are 15 — 19 days old, and the parents feed them for a further week. Sometimes the male alone feeds them, while the female prepares for a second brood.
These birds feed mainly on insects, such as locusts and beetles, and on various berries.

Bobolink

BOBOLINK
Dolichonyx oryzivorus

The Bobolink inhabits the eastern and central regions of the United States and southern Canada. In autumn, it migrates to the Central American coast and the Antilles, and sometimes to the Galápagos Islands. It is 17.5 cm long, and outside the nesting season, the male's plumage resembles that of the female, which is brown with yellow-olive shades above, the feathers having black tips. Her underparts are yellow-grey, and the wings and tail are black-brown.
The Bobolink lives in large flocks in moist grasslands. Rice was widely grown in North and South Carolina earlier in the century and the Bobolink's habit of feeding in rice paddies gave rise to its scientific name as well as to its local name of rice bird.
Bobolinks remain sociable in the nesting period, living close to each other in pairs. The nest is made in a depression in the ground, most often in a meadow. It is woven from dry leaves, grass stems and other plant material and the nesting cup is lined with fine grass. The female lays 4 — 7 eggs, pale brown with russet markings, and incubates the clutch for 11 — 13 days. To begin with, the young are fed on insects and larvae, spiders and molluscs, but as they grow bigger they eat seeds. The adults feed on seeds and shoots of green plants. After nesting, families merge together to form huge flocks which raid the fields of grain. The birds are killed in large numbers by farmers. In the nesting season, the male delivers a slow, bubbling song.

EASTERN BOX TERRAPIN
Terrapene carolina

The Eastern Box Terrapin is common in the belt from south-east Canada south to Florida and Texas. It has a tall carapace about 20 cm long. The male has a red iris, and the female a brown one. This reptile dwells in grassed and bush-covered localities. In spring, shortly after coming out of hibernation, the terrapins form pairs. In June or July, or even

Eastern Box Terrapin

Chuckwalla

earlier in the south, the female excavates a depression 10 cm deep in which she lays 4 — 5 hard-shelled eggs. She buries the clutch and stamps the earth down firmly. The young hatch after 3 months, in September or October, and immediately go into hibernation without taking any food. They leave their winter shelters in the following spring. They grow very quickly for the first 5 — 6 years, increasing in length by about 2 cm a year, but later they grow only half a centimetre annually. They are fully grown by the time they are 20 years old, their life expectancy being up to 120 years.
Terrapins feed on both plant and animal food. They eat berries and fungi, and are often found in the vicinity of heron colonies, feeding on scraps of fish dropped from the nests. In the United States, they are frequently kept as pets, and are particularly useful because they catch insects.
Five other species of the genus *Terrapene* live in Mexico and the United States.

CHUCKWALLA
Sauromalus obesus

The Chuckwalla is one of the largest North American iguanas, reaching a length of up to 40 cm. It inhabits the deserts of Nevada, Utah, Arizona and California, frequenting rocky localities of the edges of treeless canyons. It feeds exclusively on plant material. The female lays her clutch of 8 — 10 eggs in sandy soil beneath stones. This shelter provides a constant temperature of around 32° C. The eggs are about 2.5 cm long and have leathery shells. The young hatch after 50 days and are about 7.5 cm long.

Collared Lizard

COLLARED LIZARD
Crotaphytus collaris

The Collared Lizard is found in rocky deserts in the south and south-west of the United States and Mexico. It is 30 cm long and has a characteristically large head and long tail. It is a fast runner, lifting its tail obliquely upwards when on the move. It can also run on its hindlimbs, with the front of its body raised in the air. This species largely dwells on stony slopes where it feeds mainly on insects and small lizards, but sometimes on smaller representatives of its own species. It also eats berries and green shoots.
The female lays 2 — 4 leathery-shelled eggs under large stones. The young hatch after a month.

TEXAS HORNED LIZARD
Phrynosoma cornutum

The head of the Texas Horned Lizard is ornamented with two large horns, and its flanks are covered with a double row of triangular spines. It is about 15 cm long and inhabits sandy localities in Texas and northern Mexico. It can rapidly bury itself in the sand by means of zigzag movements of its body. It puts its head into the sand, pushing with its head spines, and then it digs with its side spines, throwing sand up over its back. It finishes up with whirling motions of its hindlegs and tail, disappearing within a few seconds. The nostrils are equipped with round leathery lids which can be tightly closed, enabling the lizard to sleep in the sand.
The Texas Horned Lizard feeds mainly on ants, but sometimes catches spiders with its thick, sticky tongue. The female lays 20 — 30 eggs with membranous shells in a hole in the ground. The young hatch after 4 weeks.
Twenty other species of this genus occur in Mexico

Texas Horned Lizard

and in the west of the United States. Most of them are ovoviviparous, producing 12 or so eggs which hatch in the body of the female.

Gila Monster

Banded Gecko

GILA MONSTER
Heloderma suspectum

The Gila Monster is a venomous lizard which inhabits the deserts of Utah, Nevada, Arizona and northern Mexico. It reaches a length of 60 cm. Although it lives in a dry environment, it likes to bathe. During the day, it hides in holes dug out under bushes or trees. It comes out in the evening, and slowly and seemingly clumsily, combs the neighbourhood catching insects, worms and small vertebrates. It also takes the eggs of birds and reptiles. In late June the female lays 3 — 15 eggs, and the young hatch out after 30 days.
Unlike other poisonous snakes which have venom organs in the upper jaw, the Gila Monster has venom glands in its lower jaw. The poison flows through a groove between the lip and the lower jaw, mixes with saliva, and passes through canals in the fangs to penetrate the body of the prey. The poison is relatively weak, so the snake has to hold the prey fast until the venom gets into its blood stream. Smaller animals are killed quickly but larger ones need a bigger dose. The bite is not normally lethal to man.
The Gila Monster is well adapted to its desert environment. It is active at night, coming out during the day only in the rainy summer season, for when it is exposed to direct sunshine, it can die within 13 minutes. During periods of very hot weather it takes no food, but shelters in deep holes and lives on fat stored in its tail.

BANDED GECKO
Coleonyx variegatus

The Banded Gecko is distributed in California, Arizona, Texas and New Mexico. It measures 12 — 15 cm and frequents deserts and treeless hills. Although it is nocturnal, it can occasionally be seen during the day, on boulders or rock plants. Otherwise it shelters under stones, in rocky crevices or in rodents' burrows. It is most active between 7 p. m. and 10 p. m. Although they are harmless, these small geckos have the reputation of being poisonous, and even more dangerous than venomous snakes.
Geckos feed on termites, insects and spiders, and often devour their own shed skins. They have only soft skin on their backs, so they are very vulnerable to attack. However the tail breaks off when seized, ena-

Collared Snake

Hog-nosed Snake

bling the gecko to make good its escape. A new tail soon grows, although it is not as well shaped as the original one. Snakes also prey on this lizard, and seek its eggs, which are laid in rocky crevices.
The clutch usually consists of 2 soft-shelled eggs which quickly harden in the air.
The Banded Gecko hibernates in rocky cavities during the winter.

COLLARED SNAKE
Diadophis amabilis

The Collared Snake is a common inhabitant of open bushy localities, semideserts and mountainous regions in Oregon and California. It reaches a length of 75 cm. The abdomen is yellow or orange with black markings, and the tail is rich orange below. This snake is largely found in gardens, where it hides under flat stones and feeds on small lizards and tree frogs. Young snakes consume only insects. When alarmed, this snake curls its tail into a spiral which it holds above the ground to show the warning orange colour underneath. This often deters the attacker.
The Collared Snake lays 3 — 4 eggs with soft leathery shells. Several females often lay their eggs in one hole to form a clutch of up to 50 eggs, the same sites being used every year. The young snakes hatch out after 50 days.

HOG-NOSED SNAKE
Heterodon contortrix

The Hog-nosed Snake is distributed throughout the east of the United States and north-eastern Mexico where it inhabits bush-covered localities and plantations. It reaches a length of 1 m, and its coloration is adapted to its particular habitat. Some specimens are black, with white spots on the upper lip.
In the back of their mouths, hog-nosed snakes have fangs on each side of the upper jaw, which they use to grasp and kill their prey. They hunt mainly for toads and frogs, and can swallow remarkably large prey. When a toad is caught it usually increases its volume by inflating itself, but this snake fights it vigorously, pushing its jaws forward until the large back fangs pierce the toad's swollen body and make it deflate. Sometimes it takes over an hour to push the huge mouthful into the oesophagus.
When in danger, the hog-nosed snake turns on its back and lies with its mouth open and its tongue hanging limply out. If it is touched it remains perfectly rigid and appears to be dead. In this way it deceives not only man, but various raptors and carnivores, which only hunt live prey, and avoid carrion. The snake can maintain this position for over half an hour, during which time the predator usually goes away.
In late July the female lays about 20 eggs which she buries in soft soil, using her broadened nose for digging. The young hatch after 50 days.

EASTERN BLACK RAT SNAKE
Elaphe obsoleta

The Eastern Black Rat Snake is distributed in the south of United States, on rocky, lightly bush-covered slopes. It is one of the largest colubrine snakes, reaching over 2.5 m in length. It climbs up trees and bushes in search of eggs or young nestlings, using the keeled scales on its underside to get a better purchase in crevices in the bark. This snake also raids

Eastern Black Rat Snake

Common Garter Snake

Common King Snake

henhouses for the eggs and hunts young rabbits and rats, killing them by strangulation.
In June the female lays about 10 eggs, 5 cm long, in a layer of leaves. The young hatch after 2 months and at first are covered with russet and chocolate spots.
This snake is said to warn people of the presence of venomous snakes, particularly rattlesnakes and copperheads, by gliding across their path. Unfortunately, this is only a superstition.

COMMON GARTER SNAKE
Thamnophis sirtalis

The Common Garter Snake reaches a length of 1 m. It is found in damp localities overgrown with scrub in the United States and southern Canada. It seeks its prey in marshes or in water, catching fishes, frogs and tadpoles. When disturbed, it exudes a foetid secretion from a gland situated near its cloaca. The female gives birth to live young, usually producing 20 young snakes, though occasionally up to 50 may be born. They feed on earthworms and small tadpoles.
There are 15 other species of the genus *Thamnophis,* having a similar way of life, in North and Central America.

COMMON KING SNAKE
Lampropeltis getulus

The Common King Snake measures up to 2 m and is not venomous. It lives in semidesert areas in the south of the United States and northern Mexico. Its coloration is variable, and several forms are distinguished. This snake is a typical strangler, suffocating its prey in the coils of its muscular body. It mainly hunts other species of snakes, particularly poisonous rattlesnakes. It is totally resistant to their venom, and wounds inflicted by their large venom fangs do not affect it. It enjoys general protection on account of this useful quality. Attempts have been made to introduce the Common King Snake to localities having a high occurrence of poisonous snakes, but the reptile adapts to new conditions only with great difficulty.
When the Common King Snake encounters another snake, it pounces fiercely on it and bites it. Then it squeezes the prey in the firm coils of its body and swallows it, usually head first. It also feeds on small lizards, mammals, birds and birds' eggs.
The female lays about 20 eggs which hatch after 2 months.

Milk Snake

Copperhead

MILK SNAKE
Lampropeltis triangulum

The Milk Snake is a king snake found in southern Canada and in central parts of the United States. It measures up to over 1 m in length, and obtained its name from the mistaken belief that it sucks milk from the cows at night. This snake lives under stones or among the roots of trees and bushes, and hunts small rodents, snakes and lizards in the evening.
The female lays 8 — 12 eggs in soft soil, and the young hatch after 50 — 60 days.

COPPERHEAD
Agkistrodon contortrix

The Copperhead is one of the best-known and most feared of the venomous snakes. It reaches a length of up to 120 cm and inhabits the south of the United States. It occurs in 4 subspecies, each having a characteristic coloration and pattern. These snakes are abundant only in a few localities where they are not disturbed by man. Such places are marked with signs warning of the danger of being bitten by a Copperhead. These are sluggish snakes. When they are disturbed, they merely roll into a ball and prepare to defend themselves, and they bite only when stepped upon. The bite is not usually fatal because the venom fangs are small and the poison is not very powerful. However it can be lethal for a child.
Copperheads are mountain dwellers, being found on grassy slopes, in thin woodland, and in fields and meadows, though they always stay close to the forests. They are predominantly nocturnal, but in overcast weather they may be active in the daytime. In August or September, the female bears 6 — 9 live young. In late November, copperheads seek shelter in caves where they hibernate until April.
They feed mainly on small rodents, birds, frogs, and lizards, and young copperheads also catch insects.

EASTERN DIAMOND-BACKED RATTLESNAKE
Crotalus adamanteus

The Eastern Diamond-backed Rattlesnake is one of the most venomous of the snakes. It lives in coastal regions in the south-east of North America. It reaches a length of up to 2.5 m, and it weighs 10 kg. The venom fangs are up to 3 cm long, and the venom glands store a large quantity of poison. In some areas, this rattlesnake is very abundant, especially in dry localities densely overgrown with bushes and grass, though it has been seen in water. It is nocturnal, setting out after sunset to hunt small rodents and wild rabbits. When surprised, it rolls into a ball and raises its head and tail. It shakes the tip of its tail, making loud rattling sounds to deter any intruder. It attacks only when approached very closely.
The female usually bears 8 — 12 live young in August or September. The young snakes are about 30 cm long and produce venom right from the beginning.

Eastern Diamond-backed Rattlesnake

Prairie Rattlesnake

Arboreal Salamander

Cave Salamander

PRAIRIE RATTLESNAKE
Crotalus viridis

The Prairie Rattlesnake is widespread in the southwest of the United States, especially in Arizona and California. There are several subspecies. It is up to 1.5 m long. This species makes a very loud rattling sound with the loose horny segments at the end of its tail. It feeds predominantly on small rodents. The female usually bears 10 live young.

ARBOREAL SALAMANDER
Aneides lugubris

The Arboreal Salamander is a small amphibian about 10 cm long. It occurs in open country where there are scattered groups of trees, from north-eastern Oregon to southern California, and also in New Mexico. This salamander shelters in tree hollows or under stones often in colonies of up to 30. It may also take up residence in birds' nests. It is an extremely agile animal, dexterously climbing trees up to a height of 20 m. It feeds on small insects and their larvae, and spiders, and has been seen to nibble mushrooms. An unusual feature of this salamander is that it breathes only through its skin, which is densely interwoven with arteries even extending to its finger tips.
The female lays her eggs in damp moss, or in water which has collected in a leaf, or in moist holes in fallen logs or stumps. She guards the eggs until they hatch. This salamander sometimes makes soft whistling sounds.

CAVE SALAMANDER
Eurycea lucifuga

The Cave Salamander is 12 — 13 cm long. This amphibian inhabits limestone caves in Oklahoma, Arkansas, Kentucky and Virginia. Its variable coloration ranges from yellow to orange red, patterned with black spots. It usually dwells near the entrance to a cave or in damp places under stones or overhanging rocks. It is not dependent on water, but requires a moist environment. Its lungs are not developed and it uses its skin and the lining of its mouth for breathing. It has another peculiarity in that its lower jaw is immobile, although it is jointed to the rest of the skull.
This species forages at dusk, feeding on small insects, spiders and molluscs. The female lays her eggs in damp holes and guards them until they hatch.
There are other species of this genus distributed in the east of the United States and in south-eastern Canada.

BLACK WIDOW SPIDER
Latrodectus mactans

The Black Widow Spider is the plague of the American countryside. For all its size — the female being 12 — 16 mm long and the male a mere 5 mm — this is a very dangerous insect. The female produces a small quantity of extremely strong venom, and

Black Widow Spider

Black and Yellow Argiope

Jumping spider Phidippus johnsoni

though she only bites when she is touched, the fact that this species has taken up residence in houses in rural areas, being found even in occupied rooms or among clothes in wardrobes, people are inevitably bitten. It also shelters in piles of wood, and in clusters of bananas. The venom contains a neurotoxin which damages nerve endings and gives rise to paralysis and acute pain. A healthy person usually recovers within a few weeks but children, delicate people and those sensitive to animal poisons may die. The bite of a Black Widow Spider is fatal in 4.3 per cent of cases (10 per cent in the tropics). An efficient antidote is made at the snake serums institute at Butantan in Brazil.

The Black Widow is distributed from southern Canada to Tierra del Fuego at the tip of South America, being found only in warm localities. The female has a glossy black abdomen covered with red spots while the male is entirely black. He does not bite and does not feed throughout his life. In the breeding season, the male approaches the female, though at some risk, for she is more than likely to attack and devour him. In this event she becomes a widow — hence her unusual name. She lays her eggs in a cocoon, hung on a strong web and camouflaged with tiny pieces of bark. The female feeds on insects and centipedes trapped in her strong irregularly-shaped web.

Purse-web Spider

JUMPING SPIDER
Phidippus johnsoni

The jumping spider *Phidippus johnsoni* does not make a web but leaps upon its prey. It is common in the Rocky Mountains, ranging westwards to the coast. The female is 13 mm long and the male measures 11 mm. This spider has excellent vision and can detect prey from a distance of 20 cm. It turns to face it, and when the prey comes into range the spider pounces, holds it down, and eats it. The spider is supported by a strong silk thread, so it can easily return to its take-off point. It feeds on flies and other small insects within its hunting ground. The male's legs, especially the first pair, are covered with hairs which have a striking metallic sheen. In the breeding season the male stands in front of the female, raising high his shining legs. Then he crawls rapidly on the ground, showing his glossy abdomen and hopping about in a nuptial dance to attract the female.

She lays her eggs in a web cocoon and guards them until the offspring hatch. The young spiders disperse after a few days.

Other members of this genus are common in North America. Several hundred species of the family Salticidae, which have similar ways of life and are most beautifully coloured, are found in North America, including the tropics.

Tail-less whip scorpion Tarantula marginemaculata

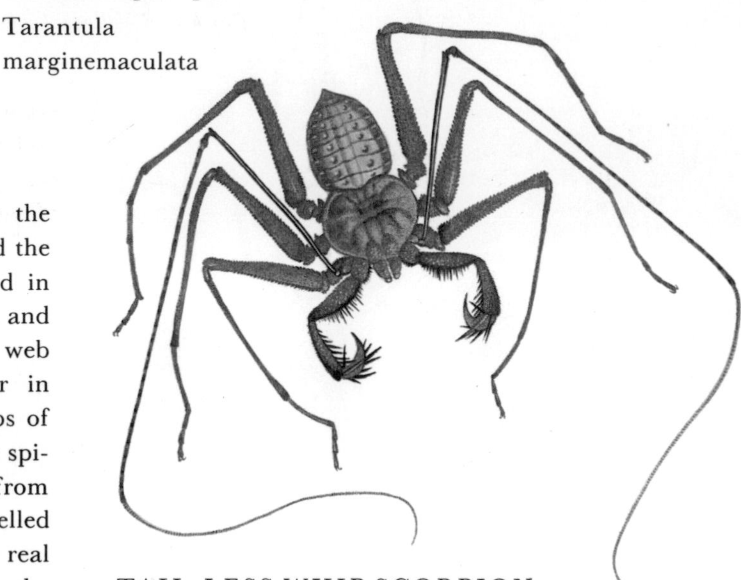

BLACK AND YELLOW ARGIOPE
Argiope aurantia

The Black and Yellow Argiope lives all over the United States and Canada. It is common around the edges of forests, in bush-covered localities, and in parks and gardens. The female is 25 mm long, and the male only 10 mm. This spider builds an orb web suspended between the branches of trees or in bushes. It covers the net with large, dense strips of web, spreading in rays from the middle. Young spiders weave only two strips, straight up and down from the centre. These strips serve to hold prey parcelled in web, and cocoons containing eggs, but their real purpose is unknown. Some zoologists believe that the spiders use up their surplus webs in them, or make them to strengthen their nets and allow them to trap larger prey.
The female lays 100 — 400 eggs in a cocoon. The young spiders hatch after a few days and disperse throughout the neighbourhood. The web mainly traps flies and wasps.

PURSE-WEB SPIDER
Atypus bicolor

The Purse-web Spider inhabits the edges of sparse woodland and clearings in the east of the United States. It is 14 mm in length and has extremely long jaws with down-turned tips and containing venom glands. This spider digs a deepish hole at the base of a tree, and weaves a silken tube inside it. The tube extends some 15 cm above the hole. This part of the structure is supported by the tree and is closed at the top. The spider spins special signalling threads along the tube, and lurks inside waiting for them to be activated. When a small insect touches the wall of the tube, the movement of the fibres announces its presence. The spider creeps unnoticed through the tunnel and stabs the insect through the tube. Then it cuts the web, pulls the prey inside, and repairs the tube again. The spider injects enzymes into the body of the prey to dissolve the tissues and speed digestion before sucking up the liquid food.
After the June rains, male spiders search the ground around the trees for partners. The female lays her eggs in a special sac in the tube. A few days later the young spiders hatch out and soon they disperse.

TAIL-LESS WHIP SCORPION
Tarantula marginemaculata

The tail-less whip scorpion *Tarantula marginemaculata* is a small arachnid, only 11 mm long. It is confined to Florida, where it inhabits the outskirts of forests, gardens and houses. It has a broad abdomen with no appendages. The first pair of legs is extremely long and whip-like. Despite its unusual appearance, this creature does not have a poisonous sting and it is quite harmless. During the day, it lives in crevices in walls, beneath bark or under stones. It comes out after sunset and mainly catches small insects. It seizes and kills prey with its palps which bear spiny protuberances.
The female lays about 30 eggs in a pouch under her abdomen. The hatched young stay with their mother for 4 — 6 days and then disperse.

WHIP-SCORPION
Mastigoproctus giganteus

The whip-scorpion *Mastigoproctus giganteus* is found in warm regions of Texas, from which it ranges to

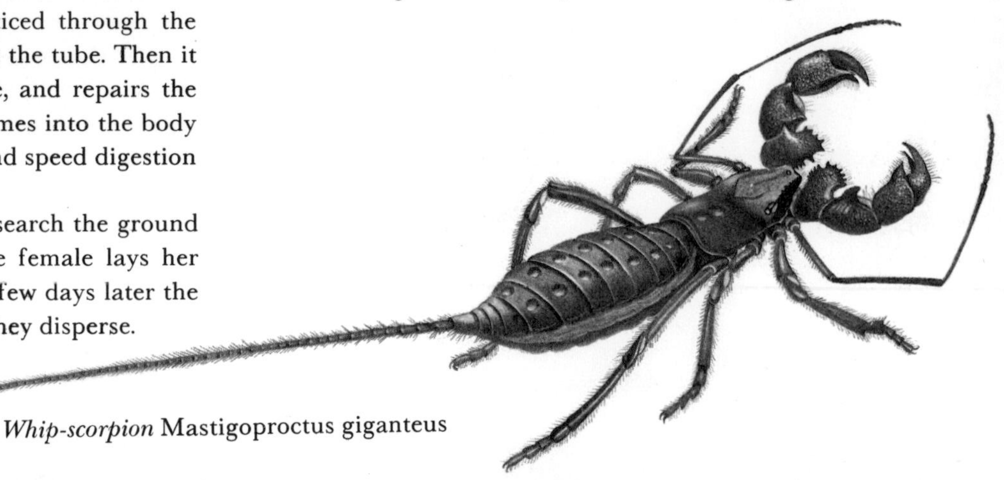

Whip-scorpion Mastigoproctus giganteus

Giant Hairy Hadrurus

Central and South America together with 70 other species of this strange arachnid. It is 15 cm long, including the tail-like appendage. Its body is flat and segmented, and the first pair of legs is not used for walking. These extremely long, thin legs are covered with cells which are sensitive to touch, and enable the whip-scorpion to find its prey. It is nocturnal, foraging after nightfall for insects, centipedes, worms and spiders. As it moves, it probes the ground with its legs until it touches the prey. Then it squeezes it, or pierces it with its spiny palps, and crushes it with its jaws. Large specimens even attack small frogs, dragging them into their burrows to eat them.
The whip-scorpion does not have a poisonous sting, and can only inflict small wounds by gripping with its palps. When threatened, it uses a different means of defence. It has a pair of glands which emit a pungent liquid containing acetic acid. This fluid damages the skin of small animals, the only effect in man is an unpleasant itch. When in danger, this whip-scorpion turns its appendage forward, and by pressing its abdomen, sprays the liquid over a distance of 30 cm, usually deterring small predatory vertebrates.
The female lays 20 — 35 eggs in a sac situated under her abdomen. The young scorpions remain on her body for a few days, but scatter after the first moult to seek damp burrows in which to live.

GIANT HAIRY HADRURUS
Hadrurus hirsutus

The Giant Hairy Hadrurus inhabits deserts and stony localities in the south-west of the United States. It reaches a length of 11 cm, which makes it one of the larger scorpions. Its coloration is variable, but its legs and tail are always bluish. During the day, it shelters in damp crevices or under large stones. It forages after sunset, catching small invertebrates with its claws. It uses the poisonous sting at the tip of its tail only in self defence.
The scorpion is solitary, seeking a mate only in the breeding season. The female bears live, whitish-coloured young, which climb on her back and attach themselves there. They begin to turn dark on the third day, and moult after a week. They move with great agility on the female's back and on the ground, feeding on prey caught by their mother. After the third moult they become independent and seek their own shelter.
Scorpions are noted for their ability to go without food for several months at a time, but they require humidity and need to drink frequently. After seizing their prey, scorpions masticate the soft parts and secrete digestive chemicals to dissolve the mixture. Then they suck in the liquid through their small mouths.

EASTERN SUBTERRANEAN TERMITE
Reticulitermes flavipes

The Eastern Subterranean Termite is a very serious pest, being one of the most destructive of some 50 termite species found in North America. It is widespread mainly in the east of the United States, but it has been accidentally introduced to many other localities. In 1952 it reached Europe, being transported by sea. It was soon breeding in large numbers and caused considerable damage to wooden buildings.
This termite forms huge colonies in intricate networks beneath the ground. A colony is established by a queen, which is about 25 mm long, and a king, which measures only 12.5 mm. After their nuptial flight, they land on a suitable spot, shed their wings and together excavate the nuptial chamber. The abdomen of the fertilized queen becomes distended and she begins to produce a continuous supply of eggs which hatch into nymphs. Large numbers of these develop into workers. They reach a length of 2.5 mm and are blind, wingless and sterile. Others, also wingless and sterile, become soldiers and defend the colony. They possess huge mandibles and measure about 5 mm. They cannot find their own food and have to be fed by the workers, which also look after

American Cockroach

Eastern Subterranean Termite

Periodical or Seventeen-Year Cicada

the queen and king and take care of the eggs and nymphs.

Termites feed on plant material both above and below the ground, particularly wood. The cellulose, which is difficult to digest, is decomposed by protozoans living in their digestive tracts. Termites maintain constant levels of temperature and humidity within the termitarium. Higher or lower humidity can prove fatal so they regulate it by the amount of saliva they exude. New queens and kings only leave the termitarium when the humidity outside is favourable.

Termites make long foraging trips, metres from the termitarium, travelling by means of tunnels built by the workers. These spread underground and even above ground on walls and girders. The subterranean passages often lead directly into wooden buildings or to telegraph poles or other structures made of timber, and as they are not visible on the surface, it is by no means unusual for a building to collapse without warning. As the Eastern Subterranean Termite lives hidden beneath the ground it is difficult to eliminate. When colonies are found they are gassed or treated with insecticides. In colder regions, termites form their colonies near district central heating systems, which provide a suitable constant temperature.

AMERICAN COCKROACH
Periplaneta americana

The American Cockroach, which measures 4 cm, is the largest of the North American cockroaches, and is very common in warm regions of North and Central America. During the day, it shelters under the bark of trees, beneath stones, or in crevices in walls and rocks. In northerly localities, it is found in houses with central heating. After dark, it crawls out and searches for food. It is omnivorous, taking any scraps of food and gnawing assorted objects, even those made of plastic. It causes most havoc in larders and food stores. In semi-dark, quiet places, it feeds during the day. It can fly for short distances, but is not a good flier. The female secretes a protective case for her eggs, which solidifies in the atmosphere. The eggs are laid inside in two rows, and hatch into wingless nymphs. At first, the female carries the case with her, but later she discards it. After hatching the nymphs moult repeatedly until they become fully-grown winged adults.

Some 50 species of cockroach are found in the United States.

PERIODICAL or SEVENTEEN-YEAR CICADA
Magicicada septendecim

The Periodical or Seventeen-year Cicada is about 6 cm long. It lives in the east of North America, in bush-covered localities and deciduous woodland, and is often found in plantations. This cicada swarms in large masses, and in some years breeds in huge numbers. In the breeding season, the female lays eggs in the bark of twigs of deciduous trees and bushes, and she and her mate then die. Although the adults live for only a short time, the nymphs undergo a record length of development. They hatch out after 6 weeks, crawl out of the twigs, and fall to the ground where they bury themselves. Here they suck sap from the roots of trees and plants, and often destroy large areas of young vegetation. They come out of the ground after the incredible time of 17 years, and climb up trees and bushes. Here they attach themselves to the bark, and after a short period, the skin on their backs bursts to release adult cicadas. In some regions in the south-east of the United States, the Seventeen-year Cicada only takes 13 years to develop.

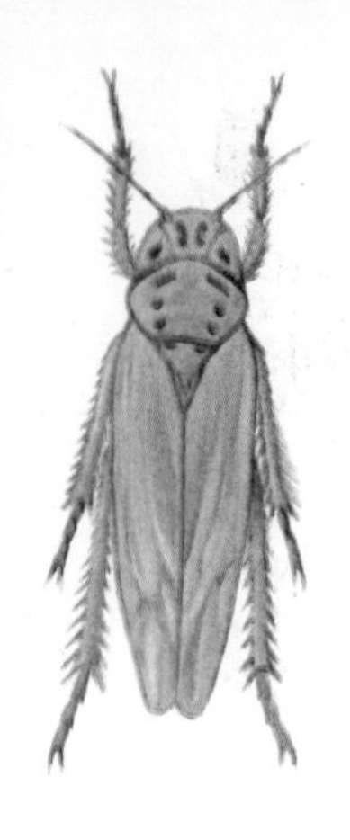

Velvet Ant

Potato Leafhopper

POTATO LEAFHOPPER
Empoasca fabae

The Potato Leafhopper is only 2.5 mm long. It is a potato pest, sucking sap from young shoots and spreading virus diseases. Though very destructive it is difficult to exterminate because it lives beneath the leaves. It occurs in the east of the Rocky Mountains, often in large numbers. It produces several broods during the year, the last generation being migratory. Its populations of several millions can spoil the potato harvest, sometimes completely destroying large fields. In autumn the insects migrate to the south of the United States, returning to the north in spring. They often fly to south and central Canada.
The female lays 2 — 3 eggs on the stems every day for 3 — 4 weeks. Nymphs hatch out after 10 days and feed on the potato leaves. They undergo metamorphosis after 2 weeks, changing into winged adults. The affected potato leaves at first develop whitish spots at edges. Later the leaves curl up and turn brown because of the toxins introduced by the sucking nymphs.

VELVET ANT
Dasymutilla occidentalis

The Velvet Ant is 2 cm long. It is found in warm regions of North America, mostly in the south-west of the United States, where it frequents dry, sandy localities. The male can fly but the wingless female moves only on the ground. She, on the other hand, possesses a sting which is absent in the male. The female lays her eggs in the nests of other hymenopterous insects, especially wasps which have the habit of storing up paralysed insects or spiders for their larvae. The Velvet Ant's larvae hatch in a few days and feed on the stores of the other wasps, later devouring their hosts' larvae. After several weeks of pupation, adult ants emerge. The males survive only until late summer. The females spend the winter rolled into balls in rock crevices or burrows, and come out in spring.

CICADA KILLER
Sphecius speciosus

The Cicada Killer occurs in warm, sandy localities in North America. It is a thread-waisted wasp about 4 cm long. It digs burrows which end in a chamber where it makes a store of food. It captures large cicadas, often bulkier than itself, paralysing them with its sting and dragging them to its nest. Sometimes it even flies with its burden. Before leaving its nest, it always closes the entrance with a piece of wood or a flat pebble. When the chamber is full of cicadas, the wasp lays an egg on them. A larva hatches out after a few days and starts to eat the paralysed cicadas. It subsequently pupates and an adult wasp emerges after a few weeks.
There are hundreds of similar species, preying on various insects or spiders in America.

ANT
Iridomyrmex humilis

Iridomyrmex humilis is a serious household pest in the warm south of the United States. This brownish-black ant is only 2.5 mm long, but its anthills are extremely extensive. Each colony has over 200 queens, which differ from the workers only in their larger size. The queens do not live a sheltered life as is the case in most species of ant, but leave the anthill regularly and wander in a large area around the nest. These ants often get on board ocean liners, concealed in pieces of merchandise, and travel to other countries where they found new colonies. In this way, *Iridomyrmex humilis* was introduced to Europe. Since so many queens are distributed through the anthills, this species is almost impossible to exterminate. In cold areas, these ants could not survive the winter, so they settle in greenhouses where they cause considerable damage to valuable crops. In tropical regions, they are particularly destructive, devouring almost anything. Huge masses of these ants devastate even the nests of birds and bees. However, they do not destroy aphids, since they feed on sweet juice exuded by these other plant pests.

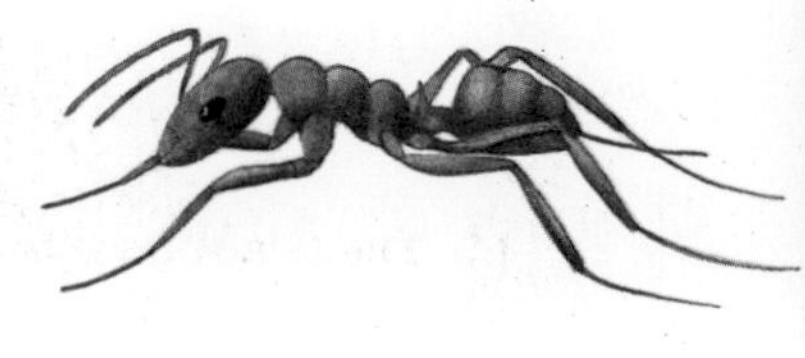

Ant
Iridomyrmex humilis

Cicada Killer

DUTCHMAN'S PIPE or BLUE SWALLOWTAIL
Battus philenor

The Dutchman's Pipe or Blue Swallowtail can be seen in summer throughout the United States except in the furthest north and north-west regions. It is most common in warm, bush-covered localities. This beautiful diurnal butterfly has a wingspan of 7.5 — 12 cm. The female lays individual orange eggs on plants of the dock family, and on Dutchman's Pipe *(Aristolochia sipho),* the hatched caterpillars feeding on the leaves. The dark brown caterpillar has long spines on the front and back part of its body, and reaches a length of about 6 cm when fully grown. At this point it attaches itself by means of a fibre to a plant stem, and pupates in a slightly oblique position, head uppermost. The pupa is pale green. The adult butterfly sucks nectar from flowers with its long proboscis.

BLACK SWALLOWTAIL
Papilio polyxenes

The Black Swallowtail is both widespread and common. It is distributed in the central and eastern regions of the United States and reaches north into southern and south-eastern Canada and south to Mexico. This butterfly has a wingspan of 7 — 9 cm, and its wing pattern is variable. The female is smaller than the male and has smaller yellow spots. Some males have only a few yellow or orange spots on the hind wings. This swallowtail lives in meadows, woodlands and gardens. It flies from flower to flower, low above the ground, sucking nectar. The female lays yellowish eggs on both wild and cultivated plants, particularly carrots, parsnips, parsley, and celery. The young caterpillars are dark brown with white saddle-shaped spots, but fully grown specimens are green with crosswise black stripes and yellow dots. The pupa is brownish, and is attached by a fibre to a plant stem, in an oblique position, with the head uppermost. In the northern area of distribution, this butterfly produces two generations, in spring and summer, while in southern regions, there are three generations from spring to autumn.

Dutchman's Pipe or Blue Swallowtail

Black Swallowtail

GIANT SWALLOWTAIL
Papilio cresphontes

The Giant Swallowtail is indeed a giant among butterflies, for it has a wingspan of 10 — 14 cm. It is distributed down the eastern side of the United States, reaching to north-eastern Mexico. It is common in the warm southern regions, producing 2 — 3 generations from May to late summer.
This swallowtail sucks nectar from flowers in gardens, fields and plantations. The female lays 400 — 500 eggs on young citrus plants and on some species of ash trees. The caterpillars are russet-coloured to orange, with large pale brown spots on the flanks and are popularly known as Orange Dogs. Like other swallowtail caterpillars, they have a pair of retractile 'horns'.

ALFALFA or ORANGE SULPHUR BUTTERFLY
Colias eurytheme

The Alfalfa or Orange Sulphur Butterfly is widespread in meadows throughout the United States, in adjacent parts of northern Canada and in southern Mexico. It has a wingspan of 4.5 — 5 cm, and the vemale is yellow. In some localities, especially near pools where butterflies drink, several hundred of them may be seen together. The female lays her eggs on alfalfa and the caterpillars feed on its leaves. Large populations of this species can cause considerable damage to alfalfa crops in agricultural regions. This butterfly has several generations each year, particularly in southern areas. The caterpillars are green, with a pale band along the body. The pupae are triangular in shape, and attach themselves to the alfalfa stems in an oblique position, head uppermost.

SPICEBUSH SWALLOWTAIL
Papilio troilus

The Spicebush Swallowtail is very common in the eastern half of the United States and in the adjacent part of Canada. It has a wingspan of 10 — 12.5 cm. The female has bluish hind wings, and the male greenish ones. It flies slowly along forest margins, above woodland paths or bush-covered slopes with solitary trees. It seeks shady and damp localities, but it also frequents fields rich in flowering plants. There are two generations in the north, and three in the south. The butterflies of the first generation, emerging from overwintering pupae in April or May, are smaller than those of the summer generation The female usually lays her eggs on spicebush, sassafras, or ash, but may also lay them on other trees. The caterpillars are green, with a longitudinal yellow stripe on the flanks and two 'eyes' on the front part of the body. There are two orange-coloured outgrowths behind the head. The rather awesome appearance of this caterpillar may help to deter predators.

ORANGE-BARRED SULPHUR
Phoebis philea

The Orange-barred Sulphur is one of the most beautiful butterflies of North America. Its occurrence is restricted to the Florida coast and north-eastern Mexico. It lives in open, bush-covered country, and has a wingspan of 6 — 8 cm. The female is a rich orange-red, with dark spots on the edges of her wings. This butterfly is common in its home range, and clouds of them sometimes fly to the American midwest. The female lays her eggs on trees of the genus *Cassia* and other related plants. The caterpillar is yellowish-green with black and yellow stripes and has small black dots along its body.

MONARCH
Danaus plexippus

The Monarch, also called Milkweed Butterfly or Wanderer, is a true traveller among butterflies. This beautiful species, having a wingspan averaging 9 cm, is widespread throughout the United States, southern Canada, and northern South America. It migrates regularly from northerly regions to Texas, California and Florida, and huge flocks can be seen resting on the trees along the way. Several hundred may alight on a single trunk, and the woodland becomes virtually covered with millions of butterflies. In spring, after hibernation, the monarchs return to their homeland in the north, and settle there in bushy localities until autumn. They produce three gener-

Giant Swallowtail

Alfalfa or Orange Sulphur Butterfly

Spicebush Swallowtail

Orange-barred Sulphur

ations in a year in the north and up to four in the south. The female lays separate eggs on the leaves and stems of milkweeds and periwinkles, and the caterpillars feed on the leaves of these plants. They are not themselves affected by poison contained in the milkweed, but the toxin accumulates in the caterpillars, and is passed on to the butterflies so both can poison other animals. A bird consuming a Monarch caterpillar suffers cramps caused by toxic glycosides, so if it recovers, it is unlikely ever to attack this kind of caterpillar again, recognizing the warning coloration of yellow, black and white crosswise stripes. An adult caterpillar is about 5 cm long. The barrel-shaped pupa is green, with several orange dots and stripes on the abdomen, and it hangs on leaf stems.

GREAT SPANGLED FRITILLARY
Speyeria cybele

The Great Spangled Fritillary has a wingspan of up to 10 cm. It is distributed throughout the United States except in the northernmost regions. It also occurs locally in southern Canada. In the north, there is only one summer generation, but in the south, there are two. These butterflies are found in open, damp meadows, alighting on the flowers of tall plants. The female lays her eggs on the leaves of violets. These provide food for the caterpillars, which feed only during the night, and hide by day in underground holes.
The caterpillars are dark brown and are covered with tall, branched spines, growing in rings. In the south, the second generation caterpillars overwinter in the

Monarch

Great Spangled Fritillary

Buckeye

Red Spotted Purple

pupal stage, and emerge in spring. The pupae are brown and hang head downwards.

BUCKEYE
Junonia coenia

The Buckeye is one of the most beautiful North American butterflies. It is distributed throughout the United States except for the north-western regions, and reaches southern Canada and northern Mexico. It is abundant in open countryside, particularly in meadows. The female lays her eggs on ribwort or other related plants. The caterpillar has rusty and whitish spots on its flanks, is black above, and is covered with spines. The pupa is predominantly silver-grey and hangs upside down on the leaves of plants. Adult butterflies hibernate in caves, cellars and similar sheltered places.

RED SPOTTED PURPLE
Limenitis astyanax

The Red Spotted Purple has a wingspan of 7.5 — 8.5 cm. It is distributed in central and eastern regions of the United States, south-western Canada and northern Mexico. It prefers open country with solitary trees and bushes. It usually flies high above the ground, but in the morning it often drinks from puddles. The female lays her eggs on wild cherries and the caterpillars feed on the leaves. Here they overwinter, and then pupate in the spring of the following year.

CAROLINA SPHINX
Manduca sexta

The Carolina Sphinx occurs almost all over the United States. It is a nocturnal moth and lives in open country, seeking the flowers of various plants and bushes, and sucking nectar. It has a wingspan of 9 — 12.5 cm, and is an excellent flier, coming out at dusk and covering fifty or so kilometres in one night. The female lays her eggs on tobacco and tomato plants and the caterpillars feed on the leaves and the great fruits. The caterpillar is green, and has seven oblique white bands along its body, each with a russet rim on top and a red, black-edged dot below. It has a characteristic pointed red horn on its abdomen, and for this reason is called the Tobacco Hornworm. When it reaches a length of about 10 cm, the caterpillar burrows into the ground and pupates in a small chamber there.

Carolina Sphinx

Moth Pholus achemon

Cecropia or Robin Moth

MOTH
Pholus achemon

Pholus achemon is a beautifully coloured nocturnal moth. It inhabits the whole of the United States, and reaches south-eastern Canada and northern Mexico. It has a wingspan of 7.5 — 10 cm. On warm nights, it visits vineyards, and the female lays her eggs on the vine leaves. She also seeks plants of black nightshade growing in open localities. The caterpillar is green or reddish-brown, with a row of five large, round spots along its flanks. The spots are whitish, black-rimmed, and with a bluish dot in the middle. The caterpillar pupates in an underground chamber.

CECROPIA or ROBIN MOTH
Hyalophora cecropia

The Cecropia or Robin Moth is one of the largest and most beautiful North American moths. It has a wingspan of about 10 — 15.5 cm and is found throughout the eastern half of the United States, to south-eastern Canada. At night, it flies in fruit plantations, gardens, and localities having a sparse cover of maples and willows. It produces only one generation annually. The female lays 200 — 300 whitish, oval eggs on the lower side of the leaves of various trees, such as cherry, plum, apple, pear, birch, maple and willow. The caterpillar reaches a length of about 10 cm and feeds on leaves. It is pale green, and has paired spiny outgrowths on its back and protuberances along its body. The first three pairs of outgrowths are huge, have globular tips, and are densely covered with bristles. On the sides of the body there are small, whitish, black-rimmed spots with pale blue dots below them. The fully-grown caterpillar pupates in a pouch-like cocoon, hung in a vertical position on the branch on which the caterpillar fed. Here it remains all winter, and the moth emerges in spring.

WATERS
AND MARSHLAND
OF NORTH AMERICA

On our trip through North America in the footsteps of its fauna, we are now entering the world of water.

The wealth of water in this continent once seemed both inexhaustible and indestructible, but from thoughtlessness, greed or ignorance man has caused great hardship both for nature and for himself, and he is still facing the consequences.

The North American lakes are among the largest natural bodies of water in the world, the Great Lakes alone covering many thousands of square kilometres. They consist of a complex of lakes both large and small, often interconnected by channels and streams. Their borders are rimmed in many places with wooded slopes, down which streams of pure water flow when the snow melts in spring. Over the last decade the crystal-clear lakes have greatly suffered from the rapid development of industry. This has polluted the water and threatened both flora and fauna. An act to protect the water and strict implementation of rules for industrial complexes halted pollution, but another danger loomed. Water from adjacent agricultural areas began to seep into some lakes. It contained artificial, nitrogen-based fertilizers which encouraged the growth of aquatic vegetation. In autumn, when the plants died off and decayed, the water became so full of rotting material that natural processes could no longer clear it. This imbalance resulted in the death of many lake animals, especially fishes and water birds, and the American administration had to allocate large resources to save the Great Lakes. Deep ditches were dug out around Lake Ontario and filled with activated carbon, to prevent infiltration by the chemicals. Many lakes have been saved, and their vegetation and fauna have returned to the original state of biological balance.

The largest lakes are found in the east of North America, on the boundary between the United States and Canada. In the area around the Great Lakes was once the home of many

Indian tribes which lived by hunting fish and waterfowl. Lake Superior is the largest lake, covering 82 400 square kilometres and reaching a depth of 307 m. Next comes Lake Huron, 59 510 square kilometres in area and 229 m deep, and Lake Michigan, 265 m deep and stretching over 58 140 square kilometres. All three lakes are in the state of Michigan, and they are interconnected. Great Bear Lake is the largest of the Canadian Lakes. It is situated in the north of Canada where it covers 31 100 square kilometres, and reaches a depth of 137 m. Great Slave Lake is somewhat smaller with an area of 28 900 square kilometres and a depth of 140 m.

The North American lakes abound in fishes of many kinds, and attract anglers from far and wide. Sea trout and salmon are the most popular game fishes. They swim from the sea to spawn in lakes near the Pacific coast. In the United States salmon spawn in hatcheries and the fry is released into the rivers and lakes. Whitefishes, especially *Coregonus clupeaformis,* abound in the Canadian lakes, as do pike and catfishes.

The North American rivers are also rich in fishes, both in terms of species and high populations. The paradise of game fishermen are the unpolluted, fast northern streams, where the Atlantic Salmon *(Salmo salar),* and Pacific species, such as the Silver Salmon *(Oncorhynchus kisutch),* spawn regularly. These streams are also visited by the Chinook Salmon *(Oncorhynchus tschawytscha).* The Rainbow Trout *(Salmo gairdneri)* occurs in fast-flowing rivers and in lakes, and is found in several resident forms. The American Brook Trout *(Salvelinus fontinalis)* is common in the cold waters of Labrador.

The Mississippi River, Father of Waters, is the longest river in North America, and has the largest basin. It rises in the vicinity of the Great Lakes, in Minnesota and Wisconsin. Its chief tributary, the Missouri, rises on the slopes of the Rocky Mountains and receives water from the glaciers. It is sometimes regarded as the main headstream of the Mississippi, which flows into the Gulf of Mexico near New Orleans in Louisiana. Its delta is covered with extensive, often densely overgrown marshes.

The Mackenzie River in Canada is the second largest river of North America. It rises in the Province of Alberta as the Athabasca, and feeds the lake of the same name. It then flows out as the Slave River, and is received by Great Slave Lake. It continues as the Mackenzie, and flows to the Beaufort Sea in the north of Canada. Other large rivers include the Winnipeg, Yucon, Columbia, Saint Lawrence and Colorado.

Large parts of North America are still covered by original marshland, and some of these areas have been declared national parks or reserves. One such is the Everglades National Park in Florida, which is the third largest protected territory in the United States, and covers 560 212 ha. Many water and wading birds thrive in the subtropical climate, as do typical plants, such as mangroves, swamp palms, water lilies and wild coffee plants. The water from Lake Okeechobee flows through the park, forming swamps and lakes with islands, and the sea brings in brackish water. The amount of water in the Everglades recently began to fall, and measures had to be taken to maintain the water level and retain the original character of the landscape. This is being done by the building of artificial dams. Alligators are one of the biggest attractions of the Everglades. These huge reptiles were once very common in the swamps and they also lived in the many rivers and lakes of the United States. Their habitats once ranged from eastern Texas across to the southern half of North Carolina and to Florida. Then swamp pools virtually swarmed with their massive bodies. Today, alligators are extremely rare. This is not to be wondered at, considering how avidly they were sought. In the decade prior to 1891, 2.5 million alligators were killed in Florida alone. In 1890, 800 specimens were captured in one small lake. The largest massacre took place in 1898. An alligator hunter called Bill Roberts was searching in an uninhabited region near Chokolskee when he found three small lakes containing so little water that the alligators were literally packed in them. Roberts and his brother set up camp beside the lakes, and killed the alligators in such numbers that, after a month, they had ten thousand alligator skins, each worth 90 cents. Over the years millions of alligators have been slaughtered in the southern states, and although they are now protected by law the region is still infested with poachers, called wild hunters. Game and law enforcement officers cannot guard every inch of this enormous territory, and modern-day poachers use not only cars, but also motor boats and hydroplanes. In the years 1940 — 1957, half a million alligators were caught in the state of Louisiana, and by the end of 1957, it was estimated that there were only 26 000 left in the state. Biologists now have to catch alligators and re-introduce them in places where they have been exterminated, and where they can now enjoy protection.

Along the coast of South Carolina there stretches a belt of brackish swamps, 15 km wide.

This is the winter quarters of thousands of water birds, who find ample supplies of food in the salty water. Further inland, there is a mountain plateau, about 75 km wide and several hundred kilometres long, where there are freshwater swamps, small streams, backwaters and grassland. The characteristic conifer of these marshes is the Swamp Cypress *(Taxodium distichum).* This is a deciduous tree with long needles. The area between swampy and sandy localities is covered by the Longleaf Pine *(Pinus palustris).* These trees reach a height of 35 m and have tufts of long needles up to 25 cm long. Older pines have irregular, branched crowns, while young pines are similar in shape to palm trees. This pine yields durable high-quality timber, and the long, fine needles are used in upholstery. Vast primary forests of Swamp Cypress and Longleaf Pine range from the Virginian coast to the Gulf of Mexico.

Experiments in the cultivation of rice were carried out in some parts of this region, but without success, and areas which were cleared of trees for the purpose soon became overgrown with vegetation again.

Vast swamps of a different type occur in the Saskatchewan region of Canada. These original, partly wooded marshes are now protected as natural reserves, and they provide nesting grounds for the last few flocks of American cranes. These once bred in many similar localities but some were exterminated and others left when the marshland was reclaimed.

And now let us set sail to meet the inhabitants of the swamps, lakes and rivers, and their neighbours on the banks.

AMERICAN MINK
Mustela vison

The American Mink is especially well known for its fur. It is distributed throughout North America except in the extreme south and south-west of the United States, and is bred on a large scale on farms. It reaches a length of 50 cm and a weight of 1 kg, females being smaller. Minks always live near water, and are excellent at swimming and diving. They often seek out large colonies of muskrats, sometimes seriously depleting them. Minks also hunt other small mammals, take the nestlings of water and marsh birds, and catch fishes and amphibians. They are active only after sunset, spending the day in burrows dug out of the banks of rivers and streams. The entrance is above the surface of the water, and the burrow has a ventilating shaft. In quiet localities, minks hunt during the day.
The breeding season begins in February or March, when the solitary males seek partners. The gestation period varies between 48 and 51 days, but it may last 38 — 70 days. The female gives birth to 1 — 10 offspring in April or May. The young are born blind and gain full sight after 30 — 35 days, at which time they are also weaned, and feed on prey brought by their mother. Later they learn to hunt and are mature after 9 months.

SEA OTTER
Enhydra lutris

The Sea Otter inhabits the rocky west coast of North America from California to Alaska. It reaches a length of 1.5 m and a weight of up to 40 kg, and is excellently adapted for life in the sea. It can dive to a depth of 50 m and stay underwater for up to 5 minutes. It hunts fishes, molluscs and crustaceans, but feeds mainly on sea urchins collected on the sea bed. The otter surfaces with its prey, turns on to its back and eats while holding the food on its chest with its forefeet. It often rests in this position in the water. The otter does not usually venture far from the coast, but it can cover large distances when swimming from island to island. It is diurnal, sleeping at night on rocky islands. Here also the female bears a single young after a gestation period of 2 months. The offspring is born furred and with open eyes. It has, however, to learn to swim. When in the water, the mother holds it on her chest as she swims on her back. When she dives for food, the young otter lies on its back on the surface of the water and waits for her.
The Sea Otter is nowadays strictly protected. In the past, particularly at the end of the 19th century, over 120 000 otters a year were killed for their beautiful fur, and the species was on the verge of extinction.

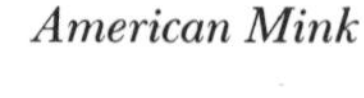

American Mink

CANADIAN OTTER
Lutra canadensis

The Canadian Otter is still common in lakes and large rivers throughout North America. It reaches a length of 130 cm including the tail and weighs about 15 kg. It is an excellent swimmer and catches fishes, small mammals, particularly muskrats, and frogs, molluscs and insects. It also takes the nestlings of water birds. It hunts skilfully in muddy water, using touch-sensitive hairs around its mouth to detect its prey.
The Canadian Otter seeks overgrown banks in which to dig out its burrow. The entrance is always underwater, so the burrow is difficult to discover. After a gestation period of 60 — 65 days, the female bears 2 — 4 young. Sometimes gestation is prolonged to 9 — 10 months. The offspring, born in late spring, are blind, gaining their sight after 28 — 35 days. At the age of 6 weeks they leave the burrow for the first time, but do not go into the water until they are 3 months old. At 4 months, they are already adept swimmers and begin to hunt. They stay with their mother for 10 — 11 months before becoming independent. They mature after 2 — 3 years.

CALIFORNIA SEA-LION
Zalophus californianus

The California Sea-lion lives off the west coast of the United States, from south-western Mexico to north-western Canada. It is most common along the Californian coast. The male measures about 2.5 m and weighs over 300 kg, but the female is much smaller. Sea-lions are fast and expert swimmers, preying on fishes and small cephalopods in the sea. They have conical teeth with which they grasp their prey. They do not chew their food, but swallow whole mouthfuls, torn off from the prey as they thrash it about in their jaws. In the stomach, the food is crushed by pebbles which the sea-lions swallow.

At the end of May, the massive males come out of the sea on to rocky islands and occupy their territories, often fighting ferociously for the best localities. Several weeks later, they are joined by the females. Each male has a harem of twenty or so females, which he guards jealously driving away rivals, and making hoarse barking sounds. He takes no food for several weeks during the rutting season, and loses a great deal of weight. In June or July, after a gestation period of 11 — 11.5 months, the female gives birth to a single cub, which weighs almost a third as much as its mother. To begin with, the cub lies on the land and the mother suckles it. Her milk contains up to 40 per cent fat, and is highly nutritious, so the offspring grows quickly. It is suckled for 6 — 8, or even 10 months. Cubs suckled for 17 months have been observed when the females did not undergo another pregnancy. A young sea-lion cannot swim and has to learn in shallow water before venturing into deeper water with its mother.

Sea Otter

Canadian Otter

California Sea-lion

Harbour Seal

HARBOUR SEAL
Phoca vitulina

The Harbour Seal is found along the west coast of North America, from northern Mexico to Alaska, and along the north-east coast of the United States. The male measures up to 2 m and weighs over 150 kg, while the rather smaller female weighs only 80 — 100 kg. This seal remains always in the same locality, wandering in its territory. When the water freezes over in its more northerly habitats, the seal keeps holes open in the ice so that it can get into the water. Young seals sometimes wander long distances up rivers in pursuit of runs of fish. In large rivers, they may travel hundreds of kilometres and may be seen even in cities.

The Harbour Seal usually lives in groups of 10 — 12. It searches for food during the day, hunting fishes, particularly plaice, and crustaceans and molluscs. It can develop a speed of up to 35 km per hour when pursuing prey, and can stay submerged for up to 15 minutes when in danger. After hunting, seals rest on sandy shores.

The female gives birth usually to a single cub after a gestation period of 340 days, the offspring being born in June or July. It is extremely active and can swim and dive. At first the mother stays close to her cub, suckles it and follows it everywhere. When twins are born, one cub dies because the female always follows just one infant. A young seal at birth weighs 10 — 15 kg and measures 90 cm, which is more than half as long as its mother. It grows quickly and is weaned at 4 — 6 weeks of age. It begins to hunt by catching crustaceans and molluscs, and later it learns to catch fish.

In recent years, the numbers of all species of seal have been declining. Annually thousands of cubs are brutally slaughtered for their skins despite public outcry at the cruelty of this method of hunting.

Canadian Beaver

CANADIAN BEAVER
Castor canadensis

The Canadian Beaver is distributed almost all over North America, but its numbers have declined, and in some localities it has been exterminated. It is a robust rodent, reaching a length of 100 cm and a weight of 30 kg. It is perfectly adapted to life in the water, having a flattened, scaly tail, webbed hind feet, and thick waterproof fur. It can close its nostrils. The Canadian Beaver lives in both still and running water, regulating the depth of water by building extensive elaborate dams about 120 cm wide and 40 — 50 cm high. They are made of branches and kept in place by small stones. Crevices are filled with mud. The dam creates an area of water deep enough not to freeze over in winter. Beavers also dig out burrows terminated by a chamber with several underwater entrances, and with vertical ventilating shafts. The burrow is lined with reeds, grass and slivers of wood. In still waters, beavers build constructions called lodges, made of branches, mud and reeds. The chamber in a lodge usually has two underwater exits. In autumn, beavers make stores of food, chiefly poplar and other twigs. In winter, they travel from their burrows along tunnels under the ice sheet to consume their stores.

Beavers feed on fresh bark, and the leaves and fruits of water and swamp vegetation. To reach the bark, beavers fell trees with a diameter of as much as 70 cm. They gnaw around the trunk with their razor-sharp teeth, selecting young poplars and birches about 10 — 20 cm across when they can.

Breeding takes place between January and March. Beavers mate in the water, and the female bears 2 — 6 young in a lined burrow after a gestation period of 105 — 107 days (sometimes 128 days). The young are born covered in fur and can see immediately. They weigh about 450 g, and they are very active and mobile. They are suckled for 2 months and then nibble fine bark and aquatic plants. Beavers mature after 2.5 years and can live to be twenty years old.

MUSKRAT
Ondatra zibethica

The Muskrat is another animal hunted for its fur. It lives in rivers, lakes and pools throughout North America, being very abundant in some parts, but having been exterminated in many others. Originally, it was hunted by American Indians for its excellent meat. The Muskrat measures 50 — 60 cm including the 20 — 25 cm-long tail, and weighs up to 1.5 kg. The almost hairless, laterally flattened tail is used as a rudder in swimming. In winter, muskrats nibble the roots of aquatic plants under the ice sheet, and from spring to autumn they feed on leaves, shoots and roots. They also catch molluscs, crustaceans and fishes, usually setting out after nightfall, and rarely feeding during the day. They either dig burrows several metres long, or build lodges, up to 1 m high and 2 m wide. These consist of heaps of pieces of aquatic plants, and have an interior chamber with several exits into the water.

After a gestation period of 28 — 30 days the female gives birth to 5 — 14 young, born blind and opening their eyes after 11 days. They are suckled for 4 weeks before becoming independent. They mature at the age of 2.5 — 3 months and are then capable of breeding. Older females may produce 4 litters in a year. This extraordinary breeding capacity has enabled muskrats to survive being hunted by man, as well as being preyed upon by raptors and carnivores. In the wild, muskrats can live to be 5 years old.

Muskrat

WIDE-NOSED MANATEE
Trichechus manatus

The Wide-nosed Manatee inhabits coastal waters off Florida, Mississippi, Louisiana, eastern Texas and Mexico. It reaches a length of 3 — 4 m and a weight of about 1 000 kg. This sluggish animal swims at

Wide-nosed Manatee

Western Grebe

Great Blue Heron

a speed of 8 km per hour, increasing it to 20 km per hour only when in danger. It is solitary and lives predominantly in shallow water both salt and brackish, sometimes entering river estuaries. It swims near the sea bed, grazing on aquatic vegetation, and using its fin-like limbs to push food toward its mouth. An adult consumes 30 — 45 kg of food daily. It feeds mainly at night, being active during the day only in places where it is undisturbed. An adult manatee can stay submerged for 12 — 15 minutes, after which time it swims to the surface to take a breath. A resting manatee lies in the water, with its body in a vertical position.

In the breeding season, males seek females and often engage in fights with their rivals. After a gestation period of 1 year, the cow bears a single cub. She gives birth underwater and pushes the calf up to the surface to take its first breath. The young manatee is about 1 m long and weighs 20 kg. It is able to swim and dive immediately after birth, but it swims to the surface every minute or so, to breathe. It stays with its mother for 2 years. Manatees have a lifespan of 12 years.

WESTERN GREBE

Aechmophorus occidentalis

The Western Grebe has a conspicuously long needle-like bill. It is found in south-western Canada and in the north-west of the United States on lakes and in marshland with a large area of water. It is about 50 cm long and has a wingspan of 100 cm. These grebes winter along the Pacific coast and on rivers in the region from Kentucky to Mexico, returning to their nesting grounds in March or April. In the courtship period, each pair of grebes performs an elaborate display. With stretched wings and necks curled in an S-shape, they run and dance on the surface of the water.

Both birds collect water plants and build a nest up to 80 cm tall in shallow water. The female lays 3 — 6 whitish eggs which turn brown during incubation. Both partners sit on the clutch for 27 days, but the female spends more time on the nest than the male. When the chicks are dry they crawl on to their parents' backs or under their wings and are carried about, although they can swim and dive. The adults collect insects for them and later find molluscs and fish fry. Adult grebes feed on small fishes, amphibians and invertebrates.

GREAT BLUE HERON
Ardea herodias

The Great Blue Heron, with a height of 130 cm and a wingspan reaching up to 2 m, is one of the largest species of herons. It occurs in regions where there are extensive lakes, rivers and swamps, throughout North and Central America. It lives in large colonies, and nests in trees or tall bushes. Herons sometimes nest on the ground on small islands. On arrival at the nesting grounds, the courtship displays begin. The males perch high on a branch, make loud screeching sounds, flutter their feathers and open their bills, to attract the females flying nearby. Each pair then builds a nest of twigs, reed stems and grass. The female lays 3 — 5 bluish eggs, which are incubated by both partners for 25 days. The young are fed by the parents, who at first thrust regurgitated food directly into their bills. Later it is placed on the edge of the nest. At the age of 8 weeks, the young leave the nest and learn to find food for themselves. They feed on small fishes, amphibians, small mammals and birds, reptiles and various invertebrates. In autumn, herons from the northerly regions migrate to warmer southern localities.

DOUBLE-CRESTED CORMORANT
Phalacrocorax auritus

The Double-crested Cormorant is the most common of the North American cormorants. It is 75 cm long and has a wingspan of 125 cm. It inhabits inland lakes from central Canada south to Florida. In winter, it travels in flocks along both the Atlantic and Pacific coast of the United States to Mexico. It nests in colonies and the pairs build nests of twigs, grass and leaves in the trees. The nest is constructed mainly by the female, while the male collects the plant material. The construction is constantly repaired and improved, and even after the young have hatched, the male continues to bring green twigs. The female lays 2 — 6 greenish eggs, and both partners take turns in incubating the clutch for 30 days. The newly hatched offspring are almost naked, and it is 3 days before they gain their sight. They are very agile, and demand food half an hour after hatching. They put their heads into their parents' throats and stimulate them to disgorge the food, usually 3 — 4 times a day. They stay on the nest for 50 days, and then jump into the water with open wings. They only begin to fly after about 55 days.

Double-crested Cormorant

The Double-crested Cormorant feeds on fish, caught at a depth of up to 20 m. Cormorants can stay under water for as long as 2 minutes. They often hunt in groups, helping each other to round up the prey.

AMERICAN WHITE PELICAN
Pelecanus erythrorhynchos

The American White Pelican nests in a zone ranging from western Canada southwards to Texas. It is common in central and southern Canada, and in the north of the United States. It measures about 145 cm

American White Pelican

and has a wingspan of 2.8 m. In the breeding season, it develops a large protuberance on the upper part of its bill. Pelicans live in colonies of up to several hundred. In autumn, they fly to the Gulf of Mexico and to the Caribbean where they spend the winter. They fly north again in April, and begin nest-building a week after their arrival. The male swims to the chosen site among reeds in shallow water with building material in his bill, and gives it to the female, who builds the nest. There are often many nests squeezed close together on a popular site. The female usually lays 2 whitish eggs, and incubation is undertaken by both partners, taking alternate shifts of 4 — 8 hours for 30 — 32 days. At first the parents regurgitate food on the nest for their young. Older chicks extract food directly from the adults' pouches. At the age of 5 weeks, young pelicans begin to swim in the neighbourhood of the nest. They start to fly at 12 weeks and become independent when they are 14 — 15 weeks old.

Pelicans feed mainly on fish, an adult consuming up to 1.5 kg every day. They hunt in groups, forming a semicircle some distance from the shore and driving the fishes into shallow water near the bank, where they catch them in the wide-stretched pouches underneath their bills. Pelicans fly with their heads held well back and their necks folded and pressed close to the body.

American or Whooping Crane

AMERICAN or WHOOPING CRANE

Grus americana

The American or Whooping Crane is 125 cm long and has a wingspan of 225 cm. It was very common on the North American continent before the arrival of the white settlers. It used to live in the vast area ranging from the Mackenzie River in Canada south to Illinois and Iowa in the United States. In autumn, their clear voices could be heard as they flew across the continent in their thousands to winter in the Gulf of Mexico. But the settlers began to reclaim the swamps where the cranes nested, and hunters killed them on a huge scale, so that by 1850, only 1400 specimens were counted in the whole of North America. In 1915 ornithologists succeeded in securing protection for this species which by now was found only in Canada. In 1923, there was only one known nesting ground, in the Canadian province of Saskat-

chewan, but egg collectors went even there, and the crane population continued to dwindle. By 1938, only 14 American cranes existed in the entire continent! At the eleventh hour, a society called the 'Cooperative Whooping Crane Project' was founded with the aim of protecting this bird. At last the cranes really began to enjoy protection even in their wintering grounds on the island of Matagorda in the Gulf of Mexico. Results were soon to be seen. In 1955 there were 28 cranes in the wild, in 1968 there were 57 of them, and their numbers go on increasing. The American Crane occupies a large territory, of about 160 ha. On arrival at the nesting grounds, the pairs carry out elaborate nuptial dances. The birds hop, run around each other, flutter their wings, and produce trumpeting sounds. They build a bulky nest of sticks and reed stems on the ground, in a thick reed bed, usually on a small island. The female lays 2 pale olive eggs with brown specks, and both partners incubate the clutch for 34 — 35 days. The chicks run and swim actively as soon as they are hatched. The parents follow them and at first feed them on regurgitated insects. Each pair usually rears only a single young, which is capable of flight in September when the cranes migrate south.
The American Crane feeds on frogs, small mammals and large insects, and on seeds, berries and green plants.

AMERICAN COOT
Fulica americana

The American Coot has an area of distribution ranging from central Canada across the United States to Mexico. It measures about 33 cm and has a wingspan of up to 65 cm. It frequents both fresh and brackish waters in coves overgrown with vegetation. In October or November it flies to the south and has been seen to walk or swim part of the way. Huge flocks spend the winter on lakes and rivers, returning to their nesting grounds in March or April. The pairs occupy territories and build nests in reed beds or clumps of swamp plants. The nest is made of stems and leaves. It often floats and has an entrance woven from long pieces of water vegetation. The female lays 6 — 9 eggs having a pale ochre ground coloration with dark purple and brownish dots and specks. Borh partners sit on the clutch for 21 — 22 days and the offspring are reared by both parents, who bring them food for the first few days after hatching. Coots feed on green plants and seeds, and in the nesting season they eat insects and other invertebrates.

American Coot

AMERICAN OYSTERCATCHER
Haematopus palliatus

The American Oystercatcher is a swamp bird about 37 cm long. It occurs on the east coast and in a few localities on the south coast of the United States, and in Mexico, living in flocks for most of the year. Oys-

American Oystercatcher

American Avocet

tercatchers have very long, thin bills, with which they probe shallow water or sandy banks for the shellfish on which they feed.

During the spring courtship period, the birds run in file or side by side, making loud trilling sounds. The pairs then occupy small nesting ranges, and fiercely defend them against intruders. The nest is usually

Ring-billed Gull

situated near water. It is a shallow depression, lined with pieces of clam shell or with sticks, leaves or pebbles. The female lays 2 — 4 pale olive, densely brown-speckled eggs, and the clutch is incubated by both partners for 26 — 28 days. After hatching, young stay in the nesting hole for up to 2 days and then accompany their parents. The adults bring them food for a few days, but the chicks soon learn to find their own prey.

AMERICAN AVOCET
Recurvirostra americana

The American Avocet is a wetland bird about 40 cm long, which inhabits almost the whole of the western half of the United States and southern Canada. It is common on lake shores and sandy islands and it also lives in vast swamps. From September to November, flocks of these birds fly to the coast in the south of the United States and Mexico, where they spend the winter, returning to the nesting grounds between March and May, depending on the locality. In the nesting season, the birds often live in colonies and make high-pitched, two-syllabic, flute-like sounds. They make their nests near the water, in muddy or sandy sites, and sometimes on patches of grass where there is low cover. The nesting cup is sparsely lined with dry grass stems, sticks or other plant material. From late April to June, and occasionally in July, the female lays 4 eggs, fawn-coloured with olive-grey and brown spots. The clutch is incubated by both partners for 24 — 25 days. The young soon leave the nest and shelter nearby among stones, clumps of grass, or in hollows.

Avocets forage mainly in shallow water, stamping on the muddy ground and flushing out tiny crustaceans and insects. They catch them with rapid flitting movements of their bills. Sometimes they catch flies, spiders or small molluscs on land. Avocets also peck at green plants and sometimes feed on seeds.

RING-BILLED GULL
Larus delawarensis

The Ring-billed Gull is native to the north-western part of the United States, central Canada and the region around the Great Lakes. It is about 40 cm long and has a wingspan of 120 cm. In the adult stage, it has yellowish-green legs and a black band round the tip of the yellow bill. It lives on lakes and in swamps containing large areas of water. For most of the year, these gulls wander in flocks throughout the United States and eastern Canada, and spend the winter on the southern coasts of the United States and Mexico. The pairs arrive at the nesting grounds in March or early April, and settle in large colonies. They build nests on small islands or in reed beds and swamps. The nest is made chiefly by the female, while the male occasionally brings her some building material. The female usually lays 3 spotted eggs, variable in coloration. She incubates the clutch alternately with the male, at intervals of 2 — 3 hours, for 25 days. The young leave the nest after 2 days and shelter in the surrounding vegetation. The parents feed them directly from bill to bill, most often fish fry, but also on insects, worms and molluscs. At the age of 3 weeks, the young gulls begin to hunt, and become independent after 5 weeks.

Bald Eagle

BALD EAGLE
Haliaeetus leucocephalus

The Bald Eagle, also called the American or White-headed Eagle, is the national bird of the United States, and appears on the national emblem. It is strictly protected. This robust raptor has a wingspan of up to 2.2 m. The male is about 80 cm long and weighs 4.5 kg. The female is larger, measuring up to 120 cm and weighing 6.5 kg. The Bald Eagle is distributed from Alaska to Florida, being found mainly beside large lakes and rivers, or on the seashore. It was once common along the coasts, but it was killed in some numbers by fishermen who regarded it as a competitor for the fish. This beautiful bird of prey has a low resistance to pesticides, which enter its body with the fish it eats. The pesticides accumulate in the eagle's tissues and cause sterility. From 50 nests observed, only 3 broods were reared. The total population of bald eagles within North America is estimated at 5 000.

The Bald Eagle forms permanent pairs, unless one of the partners perishes. The courting eagles circle high in the sky above their territory, making loud clear calls. The nest is built on a tall tree or rocky platform, or occasionally on the ground on an island. It is made of branches and lined with pine needles, grass and roots. It is up to 3 m across, and a pair uses the same nest for over 30 years, enlarging it each year. Even during incubation, the birds keep repairing it with fresh green leaves. Some nests are up to 4 m tall and weigh several hundred kilograms. They may be situated far from the water where the eagles hunt. In Florida, the female lays her eggs in December or January, while in the north nesting begins in February or March. A complete clutch consists of 2, or occasionally 3 white eggs. Both partners share incubation for 31 — 46 days, and both feed the young,

Emperor Goose

bringing large prey in their claws, and smaller prey in their beaks. They visit the nestlings 2 — 8 times a day, depending on the size of the prey. Young eagles stay in the nest for 10 — 11 weeks and then wander with their parents. On fledging, they often fly to the western coast of Alaska where food is abundant. They feed mainly on fish, but also collect carrion floating on the surface, and occasionally prey on small mammals or birds.
The Bald Eagle has an average lifespan of 40 years.

Trumpeter Swan

EMPEROR GOOSE
Anser canagicus

The Emperor Goose lives on small islands of the west coast of Canada, along the north-western coasts of the United States, in Alaska, on islands in the Bering Sea, and in north-eastern Siberia. It attains a weight of over 2.5 kg. It nests exclusively on the coasts. In autumn, small flocks of emperor geese migrate to the Aleutian, Komandorskiye and Kuril Islands, returning in spring to their nesting grounds. In Alaska they do not arrive until early June. On arrival, the geese form pairs and begin to build their nests. The partners stay together and drive away other emperor geese and other species of water birds. When attacking, the geese stretch their necks forward low above the ground, and make loud quacking noises. The nest is built on the shore above the level of the highest tide, among pieces of driftwood or on seaweed. It is made of dried plant material, and is lined with down and feathers. The female lays 3 — 8 dull white eggs, which she incubates for 24 days. During incubation, the males gather in small groups along the shore. When the young start to hatch out, the males return to their families and help to rear the broods. Adult geese begin to moult in July, and become vulnerable to carnivores, raptors and men. Eskimo hunters come to their nesting grounds, drive flocks of temporarily flightless geese to places from which they cannot escape, and easily catch them. They make warm clothing of their skins and eat the meat.
From mid-August, when the geese grow new wing feathers, the adults roam in the neighbourhood with the young ones, and they all leave the nesting grounds in late September or early October. Unlike other geese, the Emperor Goose feeds mainly on animal food — molluscs, crustaceans, sea worms and other invertebrates cast ashore, and even on the remains of fish. It also eats grass, seeds and berries.
As a result of heavy hunting, the Emperor Goose is very rare, and has become extinct in some areas. It now lives under protection in national reserves.

TRUMPETER SWAN
Cygnus buccinator

The Trumpeter Swan is the largest of the swans, reaching a length of 120 cm and having a wingspan of 2.5 m. The male weighs up to 13 kg and the female 10 kg. This swan is an inhabitant of the far north of the North American continent, and it used to be very common. Every year in spring and autumn, the swans migrated in large flocks, occupying all the northern lakes. With the first settlers came hunters who sought the swans for their beautiful plumage. Ruthless hunting exterminated the swans in many places. In the last years of the 19th century, as many as 500 swans were killed every year, with the result that, at the beginning of this century, the Trumpeter Swan was facing extinction. Only 35 specimens of this beautiful bird were counted in 1931 in the state of Montana. Just in time a reserve was established on a territory covering 155 square kilometres, and the swans came under protection. Killing was prohibited in other areas as well, and the numbers slowly started to increase. In 1954, there were 642 trumpeter swans, of which 370 lived in the Yellowstone National Park. The present-day population is estimated at 6 000, and the species will probably be saved.
The female lays 3 — 5 eggs in April or May. In the large nest, made of pieces of plant material, the clutch is enveloped in down and incubated by the female for 33 — 37 days. When the young hatch, the parents immediately take them to the water. After about 100 days, young swans are capable of flight. They weigh about 7 kg and differ from the adults in having greyish juvenile plumage. The Trumpeter Swan feeds on aquatic and land plants, seeds and berries. It also eats molluscs, insects and various other invertebrates.

CANADA GOOSE
Branta canadensis

The Canada Goose is the best-known and most widespread of the North American geese. It is found from Alaska across Canada to the eastern seaboard and in northern regions of the United States. In late autumn, large flocks migrate south to winter along the southern, western and eastern coasts of the United States and Mexico. There are 12 subspecies of the Canada Goose in the wild, largely differing in size. The subspecies *Branta canadensis minor* reaches a weight of 1.2 — 1.6 kg, while *Branta canadensis maxima* weighs up to 5 kg. For a time scientists believed that this subspecies had become extinct, but it was rediscovered in 1963 in the state of New York.
In the nesting season, the Canada Goose lives beside large lakes, along the coasts and in coves of prairie rivers. It is the first of the migrating birds to return to the north in early spring, often arriving as early as February. It returns to Alaska, in March or April. The Large V-shaped flocks contain several thousand geese, flying both by day and night. The pairs occupy territories and build nests on small islands or in swamps. The nesting cup is lined with dry leaves and grass stems. In early April, or a little later in the north, the female lays 5 — 6 eggs, covering the clutch with down. Incubation lasts 28 — 29 days, and is carried out by the female, while the male guards the nest and drives away intruders. After fledging, the families gather in flocks and wander together through the countryside. They forage on land, early in the morning and evening, feeding on grass, corn, shoots, berries and seeds, and occasionally eating insects, larvae and molluscs.

Canada Goose

Wood Duck

WOOD DUCK
Aix sponsa

The male of the Wood Duck boasts the most colourful nuptial plumage of any species of duck. However, its beauty is only temporary, for from the end of May, the males moult into a dull-coloured plumage similar to that of the females, and do not regain their courting plumage until October.
The Wood Duck is distributed through the eastern half of the United States and in the extreme north-west, where it is resident. It migrates from more northerly parts to Florida, Mexico and Jamaica. This duck is about 50 cm long. It inhabits lakes and rivers with wooded banks. It nests in tree hollows, abandoned by woodpeckers, sometimes far from water. The nest is usually 6 — 18 m above the ground. When it does not find a suitable hollow, the duck may build a nest in a shallow depression 1.5 m above the ground. The female lays 8 — 14 eggs in the bottom of the hollow. Sometimes several females lay their eggs in the same nest. The female envelops the clutch with down and incubates it for 28 — 32 days. As soon as they are dry the young jump out of the hole and drop to the ground, their fall broken by their own down, and by the grass below. The mother takes them to the water, and the family is joined by the drake. After 9 months young ducks are capable of flight. The males gain their coloration when they are 5 months old. In autumn, the families gather in large flocks which roam through the countryside. They often visit forests to find seeds and berries. For the most part these birds live on a vegetable diet, but in spring and autumn they also feed on insects, larvae, molluscs, crustaceans and worms.

BUFFLEHEAD
Bucephala albeola

The Bufflehead is a small northerly diving duck, about 35 cm long. It nests in Alaska and Canada, as far east as the province of Manitoba. In winter, it migrates south to waters in the United States and on

Bufflehead

Hooded Merganser

the Aleutian Islands. In late March or early April, the birds return to the forests and lakes of the north to nest. According to ornithological observations, the males arrive first. Two weeks later the females arrive and choose the nesting cavities. The nests are situated in tree trunks, usually 4 — 8 m above the ground. Older females often occupy the same hollow every year. The nest may occasionally be made on the ground, in a thicket or among tree roots. The female incubates her clutch of 9 — 14 brownish or greenish eggs for 22 days. When the young are dry, they jump down from the nest, and land safely in the soft grass below. The female takes them to the water where they feed on insects, tiny crustaceans and molluscs, and on water plants. Adult buffleheads also eat seeds.

HOODED MERGANSER
Lophodytes cucullatus

The Hooded Merganser is a small diving duck, which reaches a length of about 35 cm. It inhabits lakes and sluggish waters surrounded by woodland, in southern Canada and north, north-eastern, and central parts of the United States. In winter, it migrates down the west coast of the United States and down the east coast to Florida, where it lives in brackish waters. It returns to its summer habitats between mid-March and mid-April, or in May to Canada. Mergansers return either in small flocks or in pairs. The female seeks a suitable site for nesting, usually in a hole in a tree, near water. She incubates her clutch of 6 — 12 whitish eggs for 30 — 32 days. When they are dry, usually the day after hatching, the ducklings jump out of the nest and follow their mother to the water. They are capable of flying at the age of 2 months. At the end of October or in early November, families merge to form small groups and migrate south, the flocks of mergansers usually staying separate from other water birds. They feed on small fishes, crustaceans, molluscs and water insects, and they peck at plants and seeds.

Ruddy Duck

RUDDY DUCK
Oxyura jamaicensis

The Ruddy Duck is a small diving duck, about 30 cm long. It has a large area of distribution, occurring from southern Canada across the western half of the United States, through Central and South America to Tierra del Fuego. It exists in three subspecies. The North American population migrates in autumn to the south of the United States, Central America and the West Indies, where the ducks settle in river coves. They return to their nesting grounds in March or April, or in May to Canada. The Ruddy Duck frequents shallow lakes rich in aquatic and bank vegetation. The female builds her nest on a bank, directly

Belted Kingfisher

American Alligator

above water, and about 20 cm above the surface. The nest is woven from pieces of plants, and often has a roof made of stems. It is hidden in thick clumps of vegetation. The female lays 6 — 20 relatively large, white eggs and incubates them for 24 days. The male stays nearby on the water. When they are dry, the ducklings follow their mother to the water, and immediately start diving.

The Ruddy Duck feeds on small water invertebrates and on vegetable material. In September, families gather in flocks to fly to their winter grounds.

BELTED KINGFISHER
Megaceryle alcyon

The Belted Kingfisher is distributed in a zone from south-eastern Alaska across to the east of the United States, and south to Canada. It also occurs along the western coast of the United States. The northern populations winter on rivers and lakes in Central and South America. This kingfisher reaches a length of 35 cm. In the nesting season, the birds live in pairs on waters with high sand or clay banks. Here they dig out corridors, sometimes over 1 m long, terminated by a nesting chamber. They dig with their beaks, and push the material away with their feet. In May or June the female lays 5 — 8 white eggs on the bottom of the chamber. Incubation is undertaken by both partners, but the female sits most of the time, and the male feeds her. The young hatch after 21 days, and both parents feed them on tiny fishes, which also form the mainstay of the adults' diet. After 4 weeks, the offspring are fledged and leave the burrow. They stay with their parents for some time and learn to hunt. Kingfishers wait for their prey on a branch overhanging the water, or hunt in fluttering flight above the surface.

AMERICAN ALLIGATOR
Alligator mississippiensis

The American Alligator inhabits the south-eastern part of the United States, being most commonly found in the Mississippi basin. It normally reaches a length of 4 m, the largest specimen ever found being more than 6 m long. The alligator is one of the reptiles which cares for its young. The female makes a large heap of clay and leaves about 1 m tall and 2 m wide, and lays her eggs in it. Each clutch contains 20 — 70 hard-shelled eggs. During incubation, the female stays nearby and guards the clutch. After 9 — 10 weeks, the young begin to move in their shells, and hatch out three days later. When the

Florida Softshelled Turtle

female hears the sound of her offspring, she pushes away the humus to enable them to get out. The eggs are only 8 cm long, but the young alligators lie curled in them, and measure 20 — 30 cm immediately after hatching. As soon as they have left their shells, they seek shelter in the nearest shallow water, and their mother takes no further care of them. Young alligators feed on water insects and spiders, caught along the bank, and later they hunt for fishes, amphibians and various reptiles. Adult alligators feed mainly on fishes, occasionally also catching birds and small mammals. The largest specimens even take mule deer that come down to the water to drink. Alligators always flee when disturbed by man. In the breeding season, the males often engage in fierce fights and their loud, deep belowing voices can be heard far afield.

Originally, alligators had a larger area of distribution than they have now, and attempts have been made to introduce them in places where they had been exterminated. By marking individual specimens, scientists obtained fascinating data on the habits of these reptiles. When adult alligators were taken to other localities, they invariably returned to their original home. Sometimes they had to cover many kilometres and make their way through dense undergrowth. One alligator covered a distance of 13 km in 3 weeks and another travelled 32 km. Up to the age of 18 months, alligators stay near their place of birth, but then they start to wander to new grounds. Old specimens settle permanently in selected localities.

FLORIDA SOFTSHELLED TURTLE
Trionyx ferox

The Florida Softshelled Turtle inhabits waters throughout the United States except for the western and north-eastern regions. In the south, it reaches Mexico. It weighs 17.5 kg and has a carapace 45 cm long. The dorsal surface of this carapace is covered with leathery skin. This turtle has an elongated snout like a small trunk, very sharp jaws beneath fleshy lips, and three loose claws on its forelimbs. Its habitats include lakes, rivers, swamps, and localities partly covered in water. It sometimes climbs ashore and buries itself in the sand with only its head poking out, but it never goes far from the water. The Florida Softshelled Turtle can stay underwater for 20 minutes without breathing, but it prefers to lie in shallow water, buried in mud and breathing through the projecting tip of its trunk. It can stay motionless for hours, waiting for prey to approach its toothless jaws. It captures fishes, molluscs, crustaceans and insects, and large specimens take small mammals which fall into the water.

The female digs a depression in the bank and lays about 20 hard-shelled eggs in it. She buries them, moving the earth with the lower part of her carapace. The young usually hatch after 55 — 60 days, and immediately hurry to shallow water to find shelter in aquatic vegetation or mud. In the south, there are 2 — 3 clutches each year.

Fishermen often net softshelled turtles and sell them in the markets. These turtles can, however, turn dangerous and inflict serious wounds with their jaws. The meat of the Florida Softshelled Turtle is regarded as a delicacy, and its eggs are also eaten.

Alligator Snapping Turtle

ALLIGATOR SNAPPING TURTLE
Macrochelys temmincki

One of the largest freshwater turtles, the Alligator Snapping Turtle, lives in the south-east of the United States, mainly in the Mississippi and its tributaries. It is up to 1.4 m long and its carapace measures 75 cm. Large specimens weigh as much as 100 kg. This turtle has a large head which cannot be withdrawn under the carapace. It has massive jaws, the upper one being hooked like a raptor's beak. Its inconspicuous coloration makes this turtle almost invisible in shallow muddy water, where it looks like a large stone. It has many leathery outgrowths on its neck, which resemble worms and attract fishes by

Painted Turtle

their motion. Snapping turtles also lure their prey by opening their mouths and displaying their pink, worm-like tongues, which move like worms in mud. It also catches other vertebrates, and in some areas it wreaks havoc upon ducklings. Man can easily be wounded by its sharp jaws.
The Alligator Snapping Turtle lays 17 — 44 eggs in sand or humus. The clutches are avidly sought by egg-hunters, for the eggs command a high price. The young hatch after 2 months and immediately hurry to shallow water. This species is noted for its long lifespan, a specimen in the Philadelphia Zoo living to be 57 years old.

PAINTED TURTLE
Chrysemys picta

The Painted Turtle is one of the most widely distributed and handsome turtles of North America. It inhabits the whole of the central and northern part of the United States reaching to the Mississippi delta and across Texas to northern Mexico. In the north, it occurs in southern Canada. This turtle has a carapace about 20 cm long. It frequents slow-running rivers, lakes and swamps, and often crawls ashore or on to drifting logs to bask in the sun. It forages in the water, feeding on insects and larvae, molluscs and small fishes. It also nibbles green plants.
Between mid-June and early July, the female lays her eggs on dry land. She digs a shallow depression near to the water, and lays 6 — 12 tough-shelled eggs. She buries the clutch and flattens the earth with her carapace. The young hatch after 2 — 3 months and hurry to the nearest shallow water. If the eggs are laid late in the season, the young turtles hibernate in them and hatch in the spring of the following year. Newly hatched turtles measure 2.5 cm. Populations found in more northerly regions usually hibernate from October to April, waking up when the temperature of the water rises above 10° C.

MUD SNAKE
Farancia abacura

The Mud Snake is a denizen of North and South Carolina, from which it ranges to Florida and Louisiana. It is 1.2 — 1.8 m long. It is inconspicuous in the wild, sheltering in burrows, under timber in woods, or in shrubberies, always near water. When grabbed, the snake pushes the end of its strong tail into its attacker's palm, and since the tail is tipped with a thorny scale about 2 cm long, it easily penetrates the skin. Many people, particularly American Indians, fear this snake, believing that the tip is poisonous. The Mud Snake hunts smooth and slimy skinned amphibians. Frogs defend themselves by inflating their bodies, but this snake deflates them by piercing them with the tip of its tail and easily swallows them. It also catches fishes.
The female lays 20 — 100 leathery-shelled eggs in a soft depression on the bank. The young hatch after 50 days and immediately disperse.
In common with other species of snakes, the Mud Snake is said to hold its tail in its mouth and to roll along in order to go faster. Though totally untrue, this belief is widely held in many parts of the world.

Mud Snake

COMMON WATER SNAKE
Natrix sipedon

The Common Water Snake is one of the best-known North American snakes. It is distributed from southern Canada to Texas and Oklahoma. Females

Common Water Snake

reach a length of 1.2 m, while males are only 1 m long. There are several subspecies, differing in their patterns. Water snakes live on river banks and beside lakes. When this snake is threatened, it secretes a fluid from a gland situated near the base of its tail, which sometimes deters the attacker. The secretion is harmless to man.
The Common Water Snake feeds on fishes and frogs. It is often seen basking in the sun on river banks, many snakes sometimes gathering in one spot.
Like other snakes of this genus found in North America, the Common Water Snake is viviparous, bearing live young. There are usually about 25 offspring, but a female has been seen to give birth to 99. Young snakes hide beneath pieces of wood or under flat stones beside the water.

WATER MOCCASIN or COTTONMOUTH
Agkistrodon piscivorus

The Water Moccasin or Cottonmouth is a thick-bodied snake which reaches a length of 1.5 m. It lives in the southern part of the United States. The young are more colourful than the adults, usually being russet with narrow brown stripes. The Water Moccasin inhabits swamps, pools, rivers and lakes, and frequents damp localities near water. In times of flood, these snakes often collect in hundreds in one locality. They feed on fishes and frogs, and sometimes catch small birds and mammals. These venomous snakes are much feared, but they rarely bite man, for they are not aggressive, and only bite in self-defence, or when stepped upon.
The female produces up to 50 live young which soon disperse in the area.

Water Moccasin or Cottonmouth

TWO-TOED AMPHIUMA or LAMPER EEL
Amphiuma means

The Two-toed Amphiuma or Lamper Eel is a tailed amphibian, which reaches a length of 1 m. Its eel-like body carries tiny limbs with 1 — 3 toes, though they are not suitable for walking. This amphibian inhabits the south-east of the United States where it lives in rivers, lagoons and wet, swampy areas. By day, it shelters in dense water vegetation, in holes under banks, or beneath the hulls of anchored ships. It comes out after dark, swimming in a snake-like manner and catching insects, crustaceans and molluscs. In times of drought, when the water level falls, this amphibian is able to move over damp land to a deeper stretch of water. The female lays strings of relatively large eggs in holes in the mud or among plant roots, and curls her long body around the clutch to guard it. The larvae hatch after 5 months.

HELLBENDER
Cryptobranchus alleghaniensis

The Hellbender lives in large, fast-flowing mountain streams in the United States, being found in a large area from the state of New York south to northern

Two-toed Amphiuma or Lamper Eel

Hellbender

Alabama, Arkansas and South Carolina, though its distribution is discontinuous. This heavily built newt reaches a length of 75 cm, and its coloration varies, ranging from spotted black and yellow to tones of red and brown. It has 4 toes on its forelimbs and 5 on the hind ones. It mainly hunts worms, crustaceans, molluscs and insects, but large specimens catch even fishes and amphibians. The female lays her eggs in a solid string, attached at one end to a boulder to prevent the stream from carrying it away. The larvae hatch after 75 days and travel upstream, catching worms and insect larvae in the mud.

GREATER SIREN
Siren lacertina

The Greater Siren inhabits shallow rivers in the south-east of the United States. It also occurs in eastern Texas, and is found further north in Illinois and Indiana, as far as Lake Michigan. This eel-like creature reaches a length of up to 1 m and has large external gills. The very short front limbs each have four toes, and there are no back limbs. In cool weather the Greater Siren emerges about once an hour to breathe, but as the water gets warmer, it has to come up to breathe every 3 — 5 minutes. Its diet consists of insects, worms and other water animals which it digs rapidly and dexterously out of the mud. Sirens occasionally climb ashore and excavate tunnels several metres long in damp soil, and shelter in them. The female lays several hundred eggs each in a gelatinous covering.

TIGER SALAMANDER
Ambystoma tigrinum

The Tiger Salamander is an axolotl. Except in the extreme west and east it is distributed throughout the United States reaching as far as southern Canada and north-eastern Mexico. It measures up to 30 cm and has an irregular pattern of spots which differ in number and shape in individual specimens. It occurs in several subspecies which develop in different ways. Populations found in the east complete their larval stage. This phenomenon, called neoteny, is then leave the water and seek damp sites on land, especially in pine forests. The western populations of Wyoming, Colorado and New Mexico, however, never complete their development. They spend their entire lives in the water, and become sexually mature and capable of reproduction while remaining in the larval stage. This phenomenon, called neoteny, is caused by lack of iodine in the lakes inhabited by axolotls. The larvae have large, branched-out gills. If the water is low in oxygen, they breathe on the surface. During their development, they first grow their front limbs, and then the hind ones. In areas where they reach the adult stage and live on land, they return to the water in the breeding season, and come out again afterwards. This species feeds on small insects and their larvae, spiders and worms.

AMERICAN BULLFROG
Rana catesbeiana

The American Bullfrog is one of the largest species of frogs. It is found throughout the United States

Greater Siren

Tiger Salamander

and having been introduced in Cuba, it reproduced there in large numbers. It reaches a length over 20 cm, and has robust legs, up to 25 cm long. As the legs are both large and palatable, this frog is extensively hunted for food. Attempts to introduce it in Europe have so far failed.
The male has large ear drums behind its eyes, and it has a loud voice which sounds like the bellowing of a bull. Bullfrogs live in swamps, pools, lakes and river coves. They are extremely voracious, hunting insects, crustaceans, molluscs and fishes, both on land and in water. Large specimens can seize a small mammal or frog, or young snake and swallow it in one gulp. The female lays up to 5 000 eggs in clusters in shallow water. When the tadpoles hatch out, they develop in the water, growing first their hind limbs, then the forelegs, and completing their development by the age of 3.5 months, when the young frogs climb ashore. In the north the larvae remain in the water throughout the winter and complete their development in the spring of the following year.

AMERICAN PADDLEFISH
Polyodon spathula

The American Paddlefish occurs in unpolluted rivers and lakes of the United States. This unusual fish reaches a length of 2 m and a weight of 90 kg. It has a smooth skin and its upper jaw is extended into a long, flattened paddle-shaped snout. It lives at some distances from the shore. Formerly very common, the American Paddlefish became rare as a result of over-fishing and the pollution of water by industry. It spawns in March in the south and in May in the north, choosing waters with gravel-covered beds, or the edges of lakes with sandy or pebble-strewn bottoms. The eggs are extremely susceptible to pollution, soon becoming mouldy, and rotting away. This paddlefish feeds on plankton. It uses its gills to filter the oxygenated water, and swallows the tiny organisms of the plankton.

American Bullfrog

GARPIKE or LONGNOSE GAR
Lepisosteus osseus

The Longnose Gar is an interesting fish of a primitive type. It reaches a length of 1.5 m, and a weight

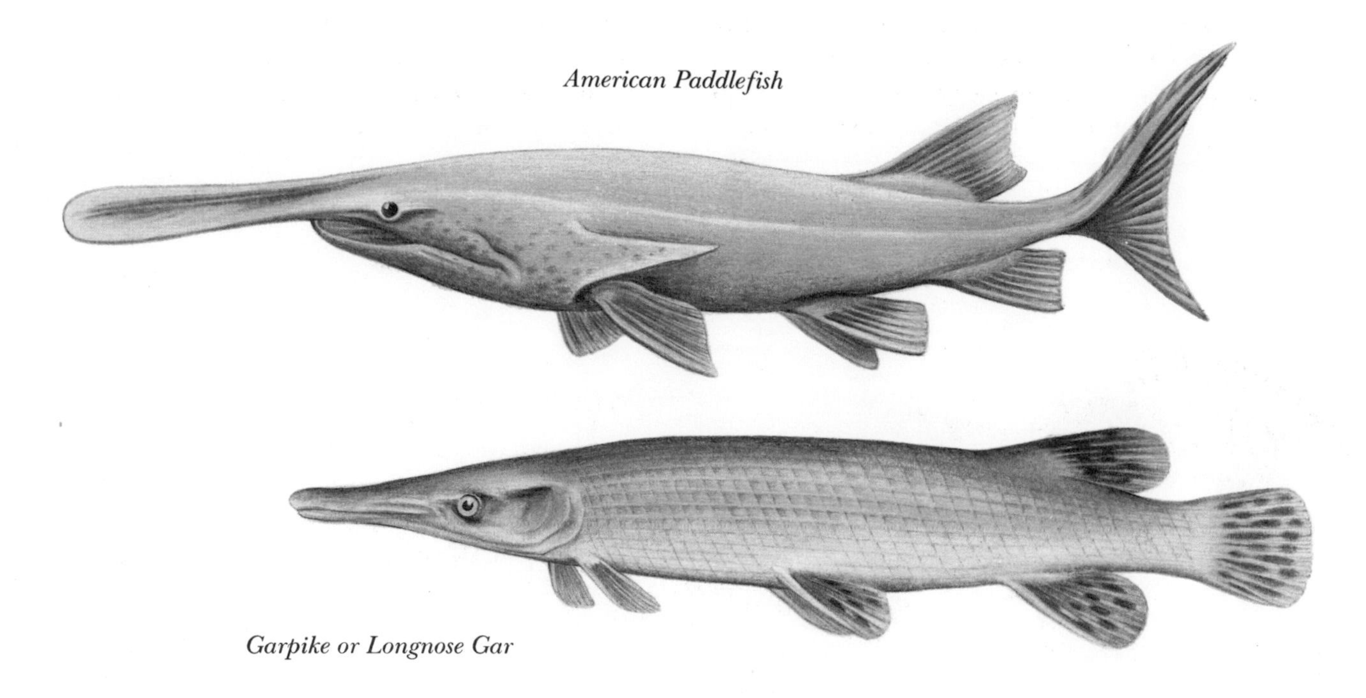

American Paddlefish

Garpike or Longnose Gar

Bowfin or Grindle

Mosquitofish, Spotted or Texas Gambusia

of 40 kg, and is characterized by a long, narrow alligator-like snout, and thick scales. It lives in the Mississippi River and its tributaries, in thick vegetation, where it preys voraciously on other fishes. Young gars feed on tiny crustaceans, worms and the larvae of insects in the plankton. Spawning takes place near the river bank, and the sticky eggs become attached to stones or plants. In winter, gars live in deep water without taking food. In summer, when the weather is hot and the water warms up quickly, they often swim to the surface to breathe.

BOWFIN or GRINDLE
Amia calva

The Bowfin, also called 'dogfish' or 'mudfish', is distributed in the Mississippi River and its tributaries, and in Lakes Huron and Erie. It reaches a length of 90 cm and a weight of 5 kg. It is nocturnal, staying near the bottom in deep water by day. At night, it swims towards the shore and pursues other fishes. Young bowfins feed on molluscs and crustaceans, worms and water insects. When the oxygen level in the water drops, bowfins breathe at the surface through a specially adapted air bladder and they can easily survive for up to 24 hours outside water. Bowfins spawn in May and June when the temperature of the water reaches 20 — 25° C. The male and female seek shallow water, where there are clumps of aquatic plants among which the male builds a nest. Here the female lays up to 70 000 eggs. The male guards the nest and later protects the fry until they are about 8 cm long. Bowfins are not in great demand as food.

MOSQUITOFISH, SPOTTED
or TEXAS GAMBUSIA
Gambusia affinis

The Mosquitofish, Spotted or Texas Gambusia inhabits American waters from the state of New Jersey to Florida, and is common in California and northern Mexico. It has also been successfully introduced to other warm parts of the world. The male reaches a length of 3.5 cm, and the female a length of 6 cm. This species is viviparous, the female producing 50 — 60 live young among thick vegetation in shallow water. The fry immediately try to hide from their own mother, who may well devour them.
Mosquitofishes feed on insects and their larvae, crustaceans and the fry of other fishes. They are particularly fond of mosquito larvae and can eat their own

Pumpkinseed Sunfish

Walleye or Blue Pike

weight in them every day. For this reason they are cultivated in regions infested by mosquitoes in order to control malaria.

PUMPKINSEED SUNFISH
Lepomis gibbosus

The Pumpkinseed Sunfish inhabits unpolluted waters of North America, from the Great Lakes to Florida. It was also introduced in Europe, where it reproduced so rapidly in some rivers such as the Danube and the Elbe, that it has become one of the less desirable species. The Pumpkinseed Sunfish is very colourful, and reaches a length of 20 cm, or even occasionally 30 cm. In the breeding season, the male digs a depression among the plants on a sandy bottom, and entices the female there to spawn. The male fertilizes the eggs, and guards them until they hatch. He then protects the fry, until they begin to disperse. This species is voracious, hunting insects, worms, crustaceans, small amphibians and the fry of other fishes.

WALLEYE or BLUE PIKE
Stizostedion vitreum

The Walleye, or Blue Pike, measures up to 1 m and weighs 9 kg. It is found in cool, deep, unpolluted lakes and rivers of North America. It also occurs in the brackish waters of river deltas. When hunting, it swims to the surface and catches smaller fishes. Walleyes are very predatory and often hunt together, rounding up their prey. Young walleyes feed on insects and small crustaceans. Spawning takes place at a depth of 3 — 5 m, and the female lays over 300 000 eggs in clusters.
In their homeland, walleyes are a popular game fish, for they are excellent to eat.

Six-spotted Fishing Spider

SIX-SPOTTED FISHING SPIDER
Dolomedes triton

The Six-spotted Fishing Spider is a small spider, distributed in the United States, east of the Rocky Mountains. The female is 20 mm long, but the male measures only 10 mm. This spider lives on the surface of the water or close to the water. It runs easily over the floating leaves of aquatic plants, and on the surface itself, for it has hairy legs which function as waterproof floats. Its prey consists mainly of insects found on water plants, but this spider sometimes waits on a leaf above the water and captures small fishes or tadpoles.
The female carries her fertilized eggs in a pouch held in her jaws. When the time for hatching approaches, she fastens several leaves together with webs, and hangs her cocoon of eggs inside this shelter. Then she settles near the nest and guards it. The young spiders leave their cocoon a week after hatching and immediately disperse.
Twelve other species of this genus are found in North America.

INDEX OF COMMON NAMES

page numbers in italics refer to illustrations

INDEX OF LATIN NAMES

TRINIDAD AND TOBAGO
ATLANTIC OCEAN
Caracas
Orinoco River
Georgetown
Paramaribo
Cayenne
Guiana Highlands
GUYANA
SURINAM
FRENCH GUYANA
VENEZUELA
Bogotá
COLOMBIA
Rio Branco
Rio Negro
Quito
Equator
Cotopaxi
Japurá R.
Manaus
Amazon River
Belém
Putumayo R.
ECUADOR
Xingú R.
Juruá R.
Huascarán
Madeira R.
Tapajós R.
Cordilleras Mountains
BRAZIL
PERU
Lima
Mato Grosso
La Paz
Nudo Coropuna
L. Titicaca
BOLIVIA
Brasilia
L. Poopó
Brazilian Highlands
PARAGUAY
Paraná R.
CHILE
Tropic of Capricorn
Asunción
São Paulo
Gran Chaco
Bermejo R.
Ojos Del Salado
ARGENTINA
PACIFIC OCEAN
URUGUAY
Aconcagua
Buenos Aires
Santiago de Chile
Colorado R.
Andes
Patagonia